TEACHING INTERCULTURAL RHETORIC AND TECHNICAL COMMUNICATION:
Theories, Curriculum, Pedagogies, and Practices

Edited by

Barry Thatcher
New Mexico State University

and

Kirk St.Amant
East Carolina University

Baywood's Technical Communications Series
Series Editor: CHARLES H. SIDES

Routledge
Taylor & Francis Group
LONDON AND NEW YORK

First published 2011 by Baywood Publishing Company, Inc.

Published 2017 by Routledge
2 Park Square, Milton Park, Abingdon, Oxon OX14 4RN
711 Third Avenue, New York, NY 10017, USA

Routledge is an imprint of the Taylor & Francis Group, an informa business

Library of Congress Catalog Number: 2009048100
ISBN 13: 978-0-89503-377-2 (hbk)

Library of Congress Cataloging-in-Publication Data

Teaching intercultural rhetoric and technical communication : theories, curriculum, pedagogies, and practices / edited by Barry Thatcher and Kirk St.Amant.
p. cm. -- (Baywood's technical communciations series)
Includes bibliographical references and index.
ISBN 978-0-89503-377-2 (cloth : alk. paper) 1. Communication of technical information--Study and teaching. 2. Technical writing--Study and teaching. 3. Intercultural communication--Study and Teaching. I. Thatcher, Barry, 1964- II. St.Amant, Kirk, 1970-
T10.5.T33 2010
601.4--dc22

2009048100

Dedication

This book is dedicated to my family, my students, and my colleagues in Mexico and Ecuador, all of whom have profoundly improved my life.

—Barry Thatcher

This book is dedicated to my daughters, Lily Catherine St.Amant and Isabelle Marie St.Amant, for being a continual source of inspiration in all that I do; to my wife, Dori St.Amant, for her unwavering patience and understanding during this project; and to the memory of Victoria Mikelonis, who has been a mentor to so many in the field of culture, communication, and education.

—Kirk St.Amant

Table of Contents

SECTION II
Curricular Perspectives: Designing and Developing Courses and Programs in Intercultural Communication

SECTION III
Connecting Instruction to Professional Practices: Merging the Workplace with the Classroom

INTRODUCTION

Thinking Globally, Teaching Locally: Understanding the Changing Nature of Technical Communication in an Age of Globalization

Kirk St.Amant

Historically, the professional practice of technical communication remained relatively restricted to Western nations. Technical communicators, in turn, tended to develop products primarily for regional or national markets (St.Amant, 2005). In these situations, technical communicators knew they would be participating in project teams composed of co-workers and subject matter experts (SMEs) who lived and worked in the same region that they did. It was also a situation in which technical communicators shared a common linguistic and cultural background with the audience for which they were producing documentation. As a result of this context, the teaching of technical communication focused on imparting knowledge and skills designed to meet the informational needs of a common cultural group (St.Amant, 2005). While this paradigm of common culture persisted for almost 50 years, the end of the 20th century brought new forces that radically shifted technical communication practices toward a more global perspective.

In the 1990s, aspects of technology and economics converged to alter the way in which individuals viewed the world and their relationship to other members of the global community. Online media, driven by the Internet and the World Wide Web, suddenly allowed persons from an array of nations to communicate easily and almost instantaneously. As a result, perceptions about space and time—particularly those related to workplace interactions—began to change. For example, the notion of the office as a physical structure with rooms and cubicles gave way to virtual workplaces in which internationally dispersed workers could collaborate on projects in real time (Burn & Barnett, 1999; Ruppel & Harrington, 2001).

Simultaneously, aspects of globalization created new opportunities for virtual production teams and the materials they created. By reducing protective barriers such as tariffs, globalization-related trends provided access to new, large markets for technical products (Beyond the digital divide, 2004; Rosenthal, 2001). As a result, undertakings that were once reserved for rare occasions, such as technical translation, became commonplace business practices. Similarly, relatively new fields such as localization and internationalization suddenly emerged as central to how organizations thought about and positioned products (Baily & Farrell, 2004; Relocating the back office, 2003; Rosenthal, 2004).

This new degree of international access also meant employees from a range of once "closed" societies could now participate in a rapidly growing global workforce. Within this framework, online media became key, for they allowed individuals in one nation to work for clients in another nation without ever having to leave home. Moreover, the kind of work that organizations could export shifted from the manufacture of physical products (e.g., athletic shoes) to the development of information products (e.g., software) and the performance of skilled service tasks (e.g., financial processing) (Relocating the back office, 2003). The result was a boom in business-process outsourcing (BPO), in which knowledge tasks, such as those associated with the development of technology products, could now be performed by globally distributed teams. And as the number of global Internet connections increased, so did access to international markets and practices associated with the outsourcing of information-based activities (Relocating the back office, 2003).

This synergy of economic forces and online media has forever changed the practice of technical communication. Today, technical communicators still think about creating informational products for their fellow countrypersons, but they also develop such products so they can be translated and localized (or internationalized) with relative speed and ease (Flint, Lord Van Slyke, Starke-Meyerring, & Thompson, 1999; Walmer, 1999; Yunker, 2002). Similarly, technical communicators still work in project teams to produce technical documentation. These teams, however, can now be composed of colleagues living in different nations and who come from a range of cultural and linguistic backgrounds.

THE SHIFT TOWARD AN INTERNATIONAL FIELD

The profession of technical communication is spreading as more non-Western nations become interested in competing in the global marketplace for technology products and informational services. Russia, for example, has a small but growing number of professionals working in technical communication (Jacobson, 2001). In India, popular newspapers tout the advantages of technical writing as a career

(Jasrotia, 2000), and China's interest in the field has led to a series of visits by delegations of technical communication teachers and practitioners affiliated with U.S. institutions, such as Marconi, Texas Tech University, and Ferris State University (Ding & Jablonski, 2001; Tegtmeier, et al., 1999; Wiles, 2003). Additionally, strong and growing chapters of the Society for Technical Communication (STC) exist in India, Israel, and Japan, and relatively newer STC chapters have been created for individuals in Slovenia, Taiwan, and Singapore (International Technical Communication Special Interest Group, 2006). These developments have also expanded the scope of what have historically been national organizations. The U.S.-based Council for Programs in Technical and Scientific Communication (CPTSC), for example, has recently worked with the Aarhus School of Business to hold CPTSC's 2009 conference in Aarhus, Denmark (University of Aarhus, 2008).

At the same time, there has been a steady international increase in the development of formal education and training programs in technical communication. For example, educational partnerships such as the Global Classroom Project have introduced an increasing number of Russian students to online instruction in technical communication (Herrington & Tretyakov, 2005). Similarly, certain faculty development programs in Ukraine and Poland have included courses in professional and technical communication in training seminars for university instructors in those nations (Mikelonis, 2000; Ukraine Business Management, 2003). In India, interest in technical communication education has prompted Calicut University, Stella Maris College, and the India Institutes of Technology (IITs) to offer a course or courses in the discipline (International Technical Communication Special Interest Group, 2005). More recently, interest in technical communication education in the People's Republic of China contributed to the development of the first Chinese-language textbook in technical communication (Duan & Gu, 2005).

THE CHALLENGE FOR TECHNICAL COMMUNICATION EDUCATORS

The increasingly international nature of the technical communication field means related educational practices must now equip students to succeed in today's globalized workplace. Such an approach requires educators to provide students with a foundation of theory and guided practices they can use to work effectively with colleagues and clients from other nations. Such instruction should also help students learn how to develop a variety of informational materials for global distribution. To achieve these objectives, educators in technical communication need to understand two key factors:

1. How cultural and linguistic aspects can affect technical communication practices on an international level

2. How educational programs in other nations prepare overseas students to work as technical communicators in the global marketplace

Addressing these objectives is no easy feat. Rather the complexities of integrating ideas on culture, language, and communication into the technical communication classroom can be daunting. It is certainly a process that would require several volumes to examine in the degree of detail needed to achieve such an objective completely. Accomplishing complex tasks, however, requires one to take an initial step forward, and the editors of this collection hope this book can serve as such an initial step others will follow.

This collection is designed to be a theory and practice book that addresses the needs of both academics and practitioners (particularly trainers) in technical communication. By examining a range of topics and approaches, the collection provides a pedagogical foundation for individuals who wish to teach, practice, or research in the area of international and intercultural communication and international education or international training. The book also provides managers and administrators with the information needed to make effective decisions related to the teaching of intercultural concepts within a technical or professional communication context.

To achieve such objectives, the entries in this collection mix theory and application to present ideas, information, teaching approaches, and professional perspectives that can help technical communication instructors address culture and discourse in a variety of educational settings. Specifically, these entries can help readers

- Learn different approaches to teaching intercultural communication concepts in various contexts
- Understand how to develop individual classes and overall curricula that focus on teaching intercultural communication
- Realize the potential for merging professional practices with educational approaches to teaching intercultural communication
- Develop effective assessment mechanisms for evaluating learning related to intercultural communication concepts and practices

Through this focus, the collection also provides readers with the range and the depth of information needed to make informed decisions about the

- Development of courses in intercultural technical communication
- Integration of intercultural communication concepts into existing technical communication classes
- Creation of curricula or programs dedicated to training students in intercultural communication

- Linking of professional practices to educational contexts related to intercultural communication
- Investing of time and resources to address intercultural communication within educational or training contexts

As such objectives are relatively new to technical communication, the chapters in this book represent original pedagogical applications and perspectives based upon different theoretical and workplace foundations. Each chapter also examines how such ideas can be applied within a range of educational contexts and can be done across a variety of cultures. This balance of theory and practice means readers from a range of backgrounds within academia (e.g., faculty, graduate students, and administrators) and industry (e.g., corporate trainers, practicing technical communicators, and managers of international organizations) can benefit from the information contained within this text

OVERVIEW OF THE TEXT

This collection contains 13 chapters organized into three thematic sections. Each section, in turn, examines a different aspect related to intercultural teaching and training. The chapters in each section also provide specific examples or ideas related to that section's particular theme.

Section I: Teaching Approaches—Introducing Intercultural Communication into Classroom Situations

The chapters in this section provide educators with approaches they can use to teach aspects of culture and communication within a dedicated intercultural communication class or in existing classes that focus on other topics (e.g., technical writing, technical editing, etc.). The chapters in this section also provide a framework for integrating educational programs across nations and cultures.

The section begins with Yvonne Cleary's "An Integrated Framework for Teaching International Communication," which introduces a holistic methodology for incorporating international issues into technical communication curricula. The primary focus of the chapter is to present a framework for immersing students and faculty in the theory *and* practice of international communication. The philosophy behind this framework is that international communication is multidisciplinary by nature. For this reason, Cleary explains, international communication education is best delivered *throughout* the curriculum, for this approach is more manageable, comprehensive, and practical than a single specialized and isolated course.

In the section's second chapter, Charles Kostelnick uses contemporary and historical examples to define the cultural aspects of human forms in technical illustrations. In his chapter, Kostelnick invites readers to consider whether human forms should be designed to diminish cultural differences among users (the modernist ideal), or whether those forms should evoke those differences (the postmodern ideal). Through this examination, Kostelnick suggests ways in which technical communication educators can introduce these issues both to teach illustration and as a vehicle for teaching intercultural communication.

Poornima Padmanabhan's chapter, the third entry in this section, provides readers with a specific example of how cultural factors can affect practice and instruction in international technical communication. To achieve this goal, Padmanabhan examines the practices of Indian technical communicators through the lens of scholarly work in cultural studies and in intercultural rhetoric. Through this perspective, Padmanabhan provides practitioners and educators in technical communication with strategies for collaborating successfully with Indian counterparts. (Such strategies are particularly timely given the growth of technical writing in and of outsourcing to India.) Educators can also use the ideas presented in this chapter to formulate new teaching approaches and concepts for preparing students to work in both globally distributed teams and in intercultural communication situations.

The fourth and final chapter in this section, Emily A. Thrush and Angela Thevenot's "Globalizing the Technical Communication Classroom: Killing Two Birds with One Stone," explores new methods of exposing students directly to other cultures. Such methods involve using readily available resources to create more feasible educational situations for technical communication instructors. To examine this idea, Thrush and Thevenot present a study in which students in an introductory technical communication course in the United States were partnered with international students in an intensive program in English as a Second Language. Together, these groups of students used a cultural-attitudes profile to explore differences, did background readings on cultural differences, planned a recommendations report for an international audience, and analyzed Web pages from their respective cultures. In presenting this approach, Thrush and Thevenot also provide readers with suggestions for implementing similar kinds of programs at their own institutions.

Section II: Curricular Perspectives—Designing and Developing Courses and Programs in Intercultural Communication

The chapters in this section present strategies for developing courses and overall curricula that focus on instruction in intercultural communication. These chapters also provide an overview of the kinds of curricular development that affects technical communication program in other nations. Through such perspectives, readers gain insights into international factors influencing technical communication practices. These insights can, in turn, help readers develop approaches for addressing such factors in an educational context.

The section begins with Shelley L. Smith and Victoria M. Mikelonis's presentation of a model developed to teach U.S. students about culture through immersion-like experiences. In their chapter, Smith and Mikelonis report the lessons learned from using this model in different training sessions. They also examine how such a model offers a conceptual and a relatively easy-to-implement framework for internationalizing technical communication courses. Additionally, the authors provide readers with texts and exercises for integrating this model into different technical communication courses.

Studying abroad is increasingly a required component of undergraduate education for U.S. students. To help students meet this requirement, colleges and universities are developing programs that are innovative both in their time frame and their content. To examine how this situation can facilitate learning about culture and communication, Deborah C. Andrews and Brent Henze's chapter analyzes the appropriateness of teaching professional writing to U.S. students in an international setting. The authors examine the logistics of such teaching in terms of how it reflects the constraints and the opportunities of an overseas context. Andrews and Henze also suggest other possibilities for incorporating an international travel segment within a professional writing course taught mainly on a U.S. campus.

James Melton's chapter shifts the discussion from offering instruction to assessing instructional effectiveness. Specifically, Melton examines how competencies in intercultural communication might be evaluated. He also discusses the use of student portfolios as a possible assessment method for such skills. In pursuing this objective, Melton examines factors such as

- Which intercultural rhetorical competencies can be assessed through portfolio evaluation and which cannot
- What criteria should be used for evaluating portfolios for intercultural competence
- How such criteria can be made comprehensible to the widest possible audience
- What kinds of portfolio documents best facilitate the assessment of intercultural competence

Through this examination, Melton looks at how such questions might be best addressed through interdisciplinary win-win collaborations involving technical communication teachers, foreign language teachers, and industry practitioners.

Dacia Dressen-Hammouda's chapter, in turn, examines how aspects of instruction and evaluation can come together in the design of overall technical communication curricula in other nations. In her chapter, "Teaching Technical Communication in France: Challenges and Prospects," Dressen-Hammouda describes the ongoing implementation of a 2-year master's degree program in intercultural technical communication at a mid-sized university in central

France (Université Blaise Pascal, Clermont 2). She begins this examination by describing the background context for the curriculum's design. Within this discussion, she explores writing instruction practices at the postsecondary level in France and reviews recent reforms that allow French universities to align their technical communication programs more closely with North American and northern European practices. Dressen-Hammouda then concludes this examination by discussing implications for creating a curriculum that raises intercultural awareness while helping French students develop necessary written communication skills. The result is a combination of insights into a particular culture, as well as ideas for teaching intercultural communication in different contexts.

The second section then concludes with Makarand (Mak) Pandit's overview of technical communication practices and related technical communication education in India. By examining the relationship between industry trends and technical communication education, Pandit's chapter provides readers with background knowledge that can help them work more effectively with Indian colleagues. The chapter also provides technical communication educators with a foundation for teaching their students about Indian technical communication practices. Through this approach, readers gain important insights into how to participate more effectively in today's global workforce.

Section III: Connecting Instruction to Professional Practices—Merging the Workplace with the Classroom

The chapters in this third and final section provide perspectives on how to integrate industry practices related to intercultural communication into various learning opportunities. The chapters in this section also examine how the industry-education relationship in different nations gave rise to certain kinds of technical communication programs in those nations. Such examinations, in turn, reveal the forces that shape technical communication education in different contexts.

Lyn F. Gattis starts this overall discussion by examining the role internships can play in teaching students about culture and communication. In her chapter, Gattis notes how one element critical to the development of praxis in an internship abroad is the ability to read contextually. That is, before a writer can fully understand an audience or produce useful written texts, "the writer must first become a 'reader,'" through a dynamic process of analyzing, interacting, and adapting (Anson & Forsberg, 1990, pp. 207–208). Using this idea, Gattis goes on to examine how interns working in international situations are better able to exercise phronesis (good judgment) about timing, rhetorical need, and the relationship of theory to application. In discussing such factors, Gattis provides practical suggestions for how educators can prepare students to become more skillful contextual readers during an intercultural internship as well as later on in their careers.

Standards, like internships, can provide a new mechanism for providing instruction on culture and communication expectations. After all, standards affect everything from the dimensions of the paper in one's printer to the number of threads on a screw. Standards also provide a form of consensus (and contention) in international technical communication. Students, however, need to understand such situations to succeed in today's global workforce; and Thomas L. Warren's chapter "ISO Standards and Cross-Cultural Communication: Materials for Teachers" presents a foundation for addressing such ideas. In his chapter, Warren provides educators with an overview of how cultural awareness affects the evolution and the eventual adoption of international standards. In examining this process, Warren explains how an international standard is developed and discusses how educators can use standards to teach students about issues in cross-cultural communication.

In contrast to the previous entries in this section, Leah Guren's chapter focuses on professional and educational practices related to technical communication in one particular nation: Israel. Through her examination of the unique challenges faced by technical communication educators in Israel, Guren overviews a number of topics, including the Israeli high tech industry and the way in which the technical communication field in Israel differs from that of North America. She also explains how such factors have affected technical communication education in Israel. Through this approach, Guren helps readers understand how technical communication education in Israel has developed to address practices that reflect the informational needs and expectations of a particular culture. Through this treatment of the topic, Guren provides readers with both insights into a particular culture and ideas for how industry can contribute to international educational practices in technical communication.

This final section concludes with Richard Draper's examination of technical communication education as it evolved in New Zealand. In his chapter, Draper explores how professional communication practices, in combination with the work of professional organizations, led to the rise of the technical communication field in New Zealand. He also discusses how these factors shaped and continue to shape related educational practices there. Through this examination, Draper provides readers with an understanding of technical communication in New Zealand and how professional educators train individuals to enter jobs in that field. Draper's chapter also provides educators with perspectives and ideas they can use to train technical communication students to work effectively with counterparts in New Zealand.

CONCLUSION

The various approaches and topics covered in these chapters provide educators and trainers with a context for understanding how the field of technical communication is becoming increasingly international and intercultural in nature.

These chapters also provide strong arguments for why it is imperative for technical communicators to learn international and intercultural communication skills. By presenting perspectives from authors who represent multiple cultures (seven nations and four continents), this collection provides readers with a truly intercultural approach to internationalizing technical communication education. Moreover, by presenting ideas, opinions, and perspectives from both academics and industry practitioners, this collection provides a well-rounded approach readers can use to examine an area where academic theory and industry practice overlap. These factors can provide educators and trainers with the foundation needed to integrate ideas on culture and communication into a variety of technical communication classes. Readers can also use these ideas and approaches to develop an overall technical communication curriculum for undergraduate majors, graduate students, or industry trainees.

REFERENCES

Anson, C. M., & Forsberg, L. L. (1990). Moving beyond the academic community: Transitional stages in professional writing. *Written Communication, 7,* 200–231.

Baily, M. N., & Farrell, D. (2004, July). Exploding the myths of offshoring. *The McKinsey Quarterly.* Retrieved November 11, 2004, from http://www.mckinseyquarterly.com/article_print.aspx?L2=7&L3=10&ar=1453

Beyond the digital divide. (2004, March 13). *The Economist: Technology Quarterly Supplement,* 8.

Burn, J., & Barnett, M. (1999). Communicating for advantage in the virtual organization. *IEEE Transactions on Professional Communication, 42,* 215–222.

Ding, D., & Jablonski, J. (2001). Challenges and opportunities: Two weeks of teaching technical communication at Suzhou University, China. *Technical Communication, 48,* 421–434.

Duan, P., & Gu, W. (2005). Technical communication and English for specific purposes: The development of technical communication in China's universities. *Technical Communication, 52,* 434–448.

Flint, P., Lord Van Slyke, M., Starke-Meyerring, D., & Thompson, A. (1999). Going online: Helping technical communicators help translators. *Technical Communication, 46,* 238–248.

Herrington, T., & Tretyakov, Y. (2005). The global classroom project: Troublemaking and troubleshooting. In K. C. Cook and K. Grant-Davie (Eds.), *Online education: Global questions, local answers* (pp. 267–283). Amityville, NY: Baywood.

International Technical Communication Special Interest Group. (2005).*Technical communication in India.* Retrieved October 3, 2005, from http://www.stcsig.org/itc/TC_worldwide.htm

International Technical Communication Special Interest Group. (2006). *Technical communication worldwide project of the ITC SIG.* Retrieved February 20, 2006, from http://www.stcsig.org/itc/TC_worldwide.htm

Jacobson, L. (2001, May). Technical writers in Russia. *Intercom,* 4–7.

Jasrotia, P. (2000). Is technical writing a viable career option? *IT People Evolve*. Retrieved October 3, 2005, from http://www.expressitpeople.com/20020805/cover.shtml

Mikelonis, V. M. (2000). Message sent versus message received: Implications for designing and training materials for central and eastern Europeans. In P. J. Hager and H. J. Scheiber (Eds.), *Managing global communication in science and technology* (pp. 205–231). New York: John Wiley and Sons.

Relocating the back office. (2003, December 11). *The Economist*. Retrieved December 20, 2003, from http://www.economist.com/displaystory.cfm?story_id=2282381

Rosenthal, B. E. (2004). META predicts offshoring will continue to grow at 20 percent clips through 2008. *Outsourcing Center*. Retrieved December 27, 2004, from http://www.outsourcing-requests.com/center/jsp/requests/print/story.jsp?id=4714

Rosenthal, B. E. (2001). Business risk. *Outsourcing Center*. Retrieved December 21, 2004, from http://www.outsourcing-requests.com/center/jsp/requests/print/story.jsp?id=1685

Ruppel, C. P., & Harrington, S. J. (2001). Sharing knowledge through intranets: A study of organizational culture and intranet implementation. *IEEE Transactions on Professional Communication, 44,* 37–52.

St.Amant, K. (2005). An online approach to teaching international outsourcing in technical communication classes. *Journal of Technical Writing and Communication, 35,* 191–201.

Tegtmeier, P. et al. (1999). China is hungry: Technical communication in the People's Republic of China. *Technical Communication, 46,* 36–41.

Ukraine Business Management Strengthening Activity Project Delivered by CEUME. (2003). *Center for Nations in Transition*. Retrieved November 30, 2005,from http://www.hhh.umn.edu/centers/cnt/programs_ceume.html

University of Aarhus. (2008). Council for Programs in Technical and Scientific Communication. Retrieved September 22, 2008, from http://www.cptsc.org/conferences/2009uofAarhus/

Walmer, D. (1999). One company's efforts to improve translation and localization. *Technical Communication, 46,* 230–237.

Wiles, D. (2003). Single sourcing and Chinese culture: A perspective on skills development within Western organizations and the People's Republic of China. *Technical Communication, 50,* 371–384.

Yunker, J. (2002). *Beyond borders: Website globalization strategies*. Indianapolis, IN: New Riders Press.

SECTION I

Teaching Approaches: Introducing Intercultural Communication into Classroom Situations

CHAPTER 1

An Integrated Framework for Teaching International Communication

Yvonne Cleary

Ireland is a small country on the cusp of Europe and a member of the European Union since 1973, yet it also has strong historical and social ties to North America. Ireland is often referred to as occupying an economic and cultural space "between Boston and Berlin." Its economy is, to a large extent, dependent on inward investment by multinational organizations, especially IT, companies.

At the University of Limerick (UL), technical communication has been a subject at the undergraduate and postgraduate levels for 14 years. Our graduates work as technical writers and localization specialists in large IT firms where they prepare documentation, software, and Web content for international markets. Several graduates also work in Continental Europe following graduation, and they often produce documentation in many languages for several locales. Most work in global corporations and have some interaction with employees in other countries. Working as a technical communication instructor here has, therefore, led me to examine ways to prepare our students for work in globalized environments and what that means today. The requirement to ensure that technical communication students are "world ready" has been especially apposite in an Irish and a European context, but in an age of globalization, outsourcing, and off-shoring, the topic has universal relevance.

While the profile of international/intercultural technical communication has increased through several initiatives in the past 10 years (for example, college courses, the Society for Technical Communication's Special Interest Group (SIG) in International Communication, recent textbooks and edited books and conferences), it remains a challenging topic for instructors. This chapter describes some of the challenges inherent in tying to "teach culture," and outlines approaches to curriculum design that tackle these challenges. These approaches are based on my experience as a technical communication instructor in an Irish university, and draw on best practices in North American and European programs.

In our programs in Ireland, two key objectives are to train students to work effectively with people from other cultures, and to train them to create effective communication products for people from other cultures. We do not teach international/intercultural communication as a unique, isolated course; rather, we incorporate these objectives into every aspect of the curriculum. This chapter presents the principles of our approach as a holistic framework for integrating international/intercultural communication into technical and professional communication programs.

THE CHALLENGES OF TEACHING INTERCULTURAL COMMUNICATION

Its emerging profile notwithstanding, intercultural technical communication is difficult to teach and moreover, to learn to apply effectively. One significant obstacle to teaching intercultural communication is the sense that the subject is not learnable. For many universities, intercultural communication is a "new" subject. Therefore, less, and often minimally useful, research/textbook material is available to inform teaching and guide learning than for established technical communication topics (such as information design or rhetoric). Hayhoe (2006) notes, "virtually all of us are completely in the dark when it comes to adapting the rhetoric and design of our own cultures for global audiences" (p. 10).

Due to the relative newness of the discipline, many guidelines are either too theoretical or too superficial to be useful. Culture and theories of culture tend to be elusive, rendering intercultural communication a delicate, often intangible, concept. Much of the literature (e.g., Hall, 1977; Hofstede, 1994; Trompenaars, 1993; Victor, 1992) examines dimensions of intercultural inquiry and spans several disciplines, including anthropology, sociology, communication studies, and business studies. It analyzes human behavior and our understanding of the world by looking at questions of perception, semiotics, space and time, power relations, and gender. Without practical examples, theory alone is likely to seem abstract and unrelated to real rhetorical situations students might encounter in their future careers. Overemphasis on theory is also likely to be off-putting for many students.

The corollary is that cultural analysis may lead us to emphasize general guidelines over specific examples, or to identify "*general* factors that influence communication, instead of investigating the *specific* contexts in which the communication occurs" (Lovitt, 1999, p. 7). Such generalization makes it difficult for students to engage with the reality of international contexts. Eagleton (2000) notes that the Romantic ideal of a homogenous culture is a myth. Differences between cultures are not straightforward and cannot be resolved effortlessly through superficial representations of a cultural group or through stereotyping. Lovitt (1999) suggests that instructors are guilty of using traditional, but not always appropriate, methods of cultural analysis to teach intercultural communication, methods that rely on broadly cataloging behavior, but "make no mention of issues that would appear central to international professional communication . . . critically important topics such as translation, localization, document design, visual communication, contrastive rhetorics, comparative genre analysis, patterns of reading and processing information, and so on" (pp. 6, 7).

De Voss, Jasken, and Hayden (2002) summarize additional problems posed by the approaches of business and technical communication textbooks to intercultural communication: assumption of an "entirely US readership"; presentation of intercultural scenarios as "problems to be dealt with" rather than new ways of seeing the world; and weak and ineffective transmission models that overlook the importance of context (p. 71). Assumption of an entirely U.S., or Western, audience is a serious yet pervasive problem. Instructors who do not include content on intercultural communication jeopardize the potential employability of their students, and ultimately reduce the attractiveness of their programs. More seriously, they implicitly reject a tenet of technical communication, respecting the audience.

AN INTEGRATED FRAMEWORK FOR TEACHING INTERCULTURAL TECHNICAL COMMUNICATION

Cultural theory is just one approach to teaching students to design information that respects cultural diversity. Ideally, it is one of multiple complementary approaches. Goswami (1999), concluding the collection *Exploring the Rhetoric of International Professional Communication: An Agenda for Teachers and Researchers*, "call[s] for the massive and challenging tasks of further developing the conceptual framework of the teaching of global professional communication and mapping this (evolving) agenda against existing teaching practices" (p. 306).

The remainder of this chapter presents the principles of our program's methodology for integrating intercultural issues into all parts of the technical communication curriculum. It presents these principles as a four-part framework:

- Integrating intercultural issues into the regular technical communication classroom
- Encouraging students to mix with other cultures
- Facilitating intercultural collaboration among faculty
- Involving industry in the intercultural communication debate.

Through these principles, students learn about intercultural issues in an integrated, not isolated, curriculum; they collaborate with peers, possibly through spending time abroad; they put theory into practice while working and studying abroad; and they forge international networks that will be important in their future careers. Faculty collaborate with colleagues worldwide and incorporate an intercultural dimension into their research. They can feed their experiences of both into the classroom. This model recognizes the complexities of intercultural communication and, through continued exposure to these complexities, the student develops the analytical and practical skills needed to work with people from varied racial, ethnic, or cultural backgrounds. Upon graduation, they are more likely to be able to identify cultural barriers, respect differences, and appreciate other cultures: a high priority, since many of our current students, regardless of where they live, will eventually work in global virtual teams. Exposure to these principles will assist them in negotiating and collaborating with colleagues in different time zones and for whom English is a second or even third language.

IMPLEMENTING THE FRAMEWORK

The process of applying the four principles (strands) described here into a program is slow and needs to be approached gradually. Some strands are already practiced in many programs, though not consistently, deliberately, or entirely. The philosophy behind this framework is that intercultural communication as, "necessarily multidisciplinary" (Hoft, 1995, p. 652), and delivered *throughout* the curriculum, is more manageable, comprehensive, and practical than through a single, specialized, and isolated course. The proposed framework can, of course, coexist in the curriculum with, and indeed supplement, a specialized intercultural technical communication course. There is some support for an integrated approach to teaching intercultural communication in the literature (see, for example, Maylath & Thrush, 2000; Starke-Meyerring, 2005).

The structure of the framework is not linear: students and faculty must constantly revisit intercultural issues. Following a period working or studying abroad, students feed their experiences back to peers and faculty during classroom discussions, thus enriching collective knowledge and reinforcing theory. The approach described here is experiential, cyclical, and constructivist.

Strand 1 involves integrating intercultural issues into the regular technical communication curriculum. This strand is based on current practice in many universities, and involves teaching intercultural issues (including aspects of culture, localization, and internationalization) within the standard technical communication curriculum, in, among other areas, rhetoric and communication theory, linguistics and language learning, research skills, document management, ethics, cultural theory, writing style, information design, legal issues, collaboration, and genres of writing. This strand is flexible and adaptable for traditional, new, and hybrid teaching methods. The primary argument in this strand is that content on culture, internationalization, translation, and globalization must appear throughout the curriculum, not only in one isolated course.

While the emergence of specialized courses in intercultural/international communication is welcome (and these courses work very well), not all institutions will have the resources or expertise to run them. Where such courses are available, they are often electives, not core curriculum subjects. So it is likely that only a small percentage of students will take them and, consequently, most students graduate without having taken any intercultural components in their program. Moreover, students need to be aware that intercultural issues themselves are not isolated.

De Voss et al. (2002) highlight improvements in how intercultural communication has been treated in textbooks since the 1960s and 1970s. Many textbooks now include information on intercultural adaptation within sections on traditional topics such as ethics, rhetoric, writing style, and graphic design (see, for example, Anderson, 2003; Lannon, 2005; Markel, 2004). Lannon discusses international/intercultural issues in his Introduction, and in sections on collaboration, audience analysis, ethics, persuasive writing, writing for a readable style, and oral presentations.

What follows is a brief discussion of how intercultural themes can be treated when covering some "traditional" topics in technical communication. The suggestions here can act as a starting point for curriculum review, with a view to mainstreaming intercultural content. Case studies, scenarios, and specific examples can inform discussion-based approaches and intercultural activities. Combined with theoretical information, they help to clarify real issues for students and bring abstract concepts (such as low/high context, speed of messages, collectivism/individualism, and masculinity/femininity) into clearer focus. Selected accessible readings with an intercultural slant to supplement existing teaching materials on these topics are presented in Appendix A.

Document management: The trend toward off shoring means that much previously indigenous work is now produced in parts of Asia and eastern Europe. During a course on document management, instructors can highlight the new realities of international environments for text production and what they imply for documentation managers: dispersed teams; different business practices; and different time zones, where team members are likely to speak different languages or at least different varieties of one language.

Ethics: Intercultural dimensions in the study of ethics in technical communication include the rhetorical insensitivity to difference and a worldwide tendency toward Westernization, which has made Western design the norm, and the manipulation of other cultures' technological or information deficits (Lannon, 2005).

Writing style: Instructors can stress how good technical writing permeates linguistic and cultural borders. Most style guidelines for technical writing apply tenfold when writing for an international audience. The mantras we repeat to students, such as "don't use the passive voice," "write clear, correct, concise language," "use terminology consistently," and "avoid humor" are rationalized in technical writing textbooks generally, but become even more relevant for text that must be used by nonnative English speakers or that must be localized.

Information, interaction, and graphic design: Much research suggests that for international audiences, replacing text with graphics and multimedia may reduce cultural barriers. Considerations for international texts include text direction; typography; choice of graphics and clip art; conventions for screen shots, color choices, and especially connotations of color combinations; and technological infrastructure of users. Examples of how aspects of information design can be culture-bound help students to make sense of abstract concepts such as cultural responses to color.

Legal issues: Legislation is not uniform across all jurisdictions, even though products and documentation (particularly online documentation) often are. Anyone developing online content, especially content to be delivered via the Internet, knows that intellectual property is a minefield, and instructors who have developed online courses in technical communication are already aware of issues such as ownership of content, data protection, plagiarism, and security. In addition, instructors need to alert students to intercultural differences in consumer product legislation, health and safety, legal liability, and accessibility for users with disabilities.

Collaboration: Collaboration has frequently been identified as an essential, even the most important, feature of technical communication programs. In our programs, we encourage collaboration through group work and, increasingly, through online communication technologies such as learning management systems, wikis, blogs, discussion forums, listservs, and online telephone software. These technologies will also facilitate collaboration with peers from other cultures in worldwide institutions, for students and faculty. In the past year, postgraduate technical communication students at the University of Limerick have participated in virtual team collaboration with students at the University of Central Florida, Orlando. This collaboration has increased their respect for another culture, their understanding of how to communicate with colleagues from another culture, and their confidence in their own abilities as communicators.

Technology: Hoft (1995) explains how some tools are more amenable than others to the production of internationalized (localizable) documents. Structured mark-up languages, especially XML, enable writers to create documents with consistent formatting, independent of platform or software. The XML-based DITA (Darwin Interchange Technology Architecture) may also be capable of "closing the globalization gap" (Rabkin, 2005).

Rhetoric and communication theory: If cultural theory is not addressed in a specialized course, topics such as global audiences, new media, globalization, and technology and society can be incorporated into a course on rhetoric/communication studies.

In addition, where possible, students should be encouraged to study a foreign language, since even a rudimentary grasp helps writers to appreciate the difficulties of translating, for example, noun strings, passive constructions, and inconsistent terminology. Exposure to a foreign language also highlights cultural differences, such as levels of formality (for example, most European languages have a polite and familiar form of address, while languages such as Japanese have several levels of formality). However, encouraging English speakers to learn a foreign language is indeed a challenge in an era of monolingualism, where English has become the *lingua franca* of business, finance, and IT. Even when technical communication students do not study a foreign language, an understanding of linguistics helps a writer to conceptualize grammar, syntax, and semantics, all key to developing a careful writing style.

Intercultural issues should also be incorporated into training on research practices. Research skills, of themselves, do not differ largely from culture to culture, but an introduction to research in technical communication is incomplete if it omits reference to research topics in intercultural communication. Potential research projects are myriad, including information design for international audiences, application of international variables, translation issues, controlled language use, ideologies of globalization, technology, and intercultural collaboration. Moreover, introducing students to the logistics of conducting research must mention issues peculiar to researching another culture, including language barriers, social mores, and logistical problems such as time difference, travel time, and costs.

Strand 2 involves encouraging students to mix with other cultures. De Voss et al. (2002) stress "encouraging students to move intercultural communication beyond the classroom" as one of the challenges to teaching intercultural communication. This strand is widely practiced throughout Europe and increasingly in North America. The most important European intercultural initiatives are university placements in other countries (placements are required for many programs). These placements enable students to experience firsthand intercultural communication problems, scenarios, and solutions, and enable them to extend their range of intercultural communication skills beyond the theory-bound approaches of the classroom. Residence abroad has been an integral part of language degree programs in Europe for many years. The European Union (EU) policy framework supports language learning and mobility. Indeed, mobility—the free movement of citizens within member states—is a central arm of EU policy, implemented within the educational sphere through the Socrates[1] program. Geographical proximity to other jurisdictions ensures that, for most European students, traveling to a different country for a semester or longer is not as daunting or difficult as it would be for a student in the United States. More

[1] The European Union (EU) Socrates student/faculty exchange programs support international and interinstitutional collaboration. More information is available at: http://ec.europa.eu/education/programmes/socrates/erasmus/erasmus_en.html

important from the practical perspective of impoverished students is the recent cheap flight phenomenon, where budget airlines offer low-cost (or sometimes free) flights between destinations throughout Europe. In fact, flights are substantially cheaper in Europe now than they were 10 years ago, making international travel a much more democratic and practicable enterprise. Students know they can fly home if they're feeling lonely or that family members can easily visit them.

There is strong evidence to suggest that international exchanges are very valuable for students. An EU Socrates evaluation study (Teichler, Gordon, & Maiworm, 2000) suggests that students use the study abroad experience to broaden their education, improve their language skills, and become integrated into the host country culture.

> [Students] mainly took
>
> - Courses on topics that are not available at the home institution (69%)
> - Language courses in the host country language (55%)
> - Courses to broaden their academic and cultural background (52%)
> - Courses involving teaching methods that are not used at the home institution (50%). (p. 84)

The report observes that about two thirds of students on Socrates exchanges engaged with the culture of the target country, listening to or reading local news, visiting museums, traveling in the host country and befriending "host country nationals . . . 62 per cent of the 1998/99 ERASMUS students considered themselves well integrated in the academic life and 66 per cent in the social life of the host country" (p. 88).

Students also recognized the value of the experience, reporting perceived improvements in foreign language proficiency, academic progress, and knowledge of the host country. Students who participated in the survey were overwhelmingly positive about the overall value of the experience.

The same report discusses the impact of the Socrates experience beyond graduation. It shows that graduates who have studied abroad are more likely to seek work with an intercultural dimension, often overseas; have a smooth transition to the workplace; and have slightly higher status in the workplace than colleagues who did not study abroad. At the University of Limerick, it has been our experience that students who study abroad go on to seek employment abroad upon graduation, often in the country they lived in as students.

Of course, many program structures do not facilitate periods spent abroad, and impecunious students cannot afford to spend extended periods studying in another jurisdiction. Furthermore, although in a European context it is feasible to require a sojourn in another country as part of a program of study, for U.S. students, such a requirement would prove much more financially severe.

While a period abroad cements understanding of other cultures and work practices, and must be encouraged, where it is not practicable, there are many ways of moving intercultural inquiry out of the classroom and into the realm of real life. Some are listed below.

Hosting an international student: Interaction with students who are on placement from universities worldwide deepens our collective understanding of intercultural communication.

Making foreign language study a priority: The EU encourages all students to learn a modern language other than their first language and has an aspirational policy of "mother tongue plus two"[2]—that students would be able to speak three languages proficiently.

Organizing online discussions and chats: This enables collaboration among students from various institutions, jurisdictions, and cultures. Use of new technologies for interinstitutional collaboration can increase students' intercultural awareness and proficiency.

Encouraging students to join international/intercultural technical communication groups such as Intecom and the STC's International Technical Communication Special Interest Group (SIG): These communities offer students insight into the types of technical communication that require consideration of other cultures. They are also an opportunity for students to network with professionals from many jurisdictions.

Exploring cultural variety in their own classroom: The phenomenon of global migration has resulted in cultural diversity. Starke-Meyerring (2005) calls this tendency "pluralized identities" or "blurred boundaries." In most classrooms, some students, and possibly instructors, are immigrants or migrants, whose varied cultural experiences can enrich collective understanding of intercultural issues. As Starke-Meyerring observes, "typical boundaries between national and international communication are often meaningless to [migrants/immigrants] because they have mixed or hybrid identities" (p. 477).

Strand 3 involves intercultural faculty collaboration in research and teaching. Faculty research into work practices abroad will also stimulate new dialogues in the classroom about intercultural issues. Some suggestions for increasing intercultural faculty are listed here.

Online faculty exchanges and discussion: New technologies have increased the potential for more online collaboration. For example, faculty can share classes online, design collaborative assignments for students, design common assignments for students, organize student and faculty exchanges, and participate in interinstitutional research projects.

[2] EU White Paper on Education (1995): http://ec.europa.eu/education/doc/official/keydoc/lb-en.pdf

Teaching-staff mobility: Within Europe, the Socrates program supports teaching-staff mobility as well as student mobility. The Socrates report from 2000 notes the following perceived impact of staff mobility: "Most mobile teachers believed that the teaching period abroad contributed to the improvement of their international and intercultural understanding, helped them to become familiar with other teaching methods, and was valuable for improving their research contacts" (p. 136). Short-term and longer exchanges are also gradually becoming popular for instructors worldwide. International programs can be developed directly with universities abroad or under the auspices of government. EU funding is available for European partnership master's degrees under the Erasmus Mundus label, which aims to "enhance quality in European higher education and to promote intercultural understanding through co-operation with third countries" (EU, 2006).

Collaborative research: Faculty research with academics worldwide enables us to put theory into practice, dealing with international time zones, language barriers, and different attitudes to work. Even when we work with international colleagues who share our mother tongue, linguistic issues such as different spelling conventions and different designations for everyday items arise. We are also exposed to different work practices, pressures, systems of academic life, and even holidays and customs.

Strand 4 involves collaboration with industry to examine international/intercultural questions, through seminars, workshops, industry-focused project work, and where possible, with graduates now working in international roles. Stark-Meyerring refers to the "globalization of labor and the transnationalization of business practices" (2005, p. 474), which have made virtual teams the norm and cross-cultural collaboration in the workplace a requirement. Since corporations need employees who can effectively work in multicultural environments, it is important to build links with industry into a program. Industry links take time to create; some ways of forging them are identified here.

Encourage students to go on work placements/internships in other countries: In many universities, undergraduate internships are compulsory; as part of their degree program, students spend between 6 and 8 months on a work placement, usually organized by their university. Only some work placements involve overseas travel, but humanities students often spend their internship working outside their own country. For language students, the internship is usually in a country where one of their target languages is spoken. Students of other disciplines, particularly of business, may also be encouraged to request work placements abroad. On international internships, students are exposed to international work practices, challenges, and rewards. They work through the obvious differences: language, time, and business practices. Through multinational companies' worldwide penetration, students on a work placement, even in their own jurisdiction, have exposure to an international work environment and work networks.

Encourage students and faculty to join SIGs and international technical communication organizations: As outlined in Strand 2, joining international organizations enables members to mix with other cultures. It will also stretch their potential to network with technical writers internationally, since membership of SIGs and professional organizations usually straddles academia and industry. Through international organizations, students and faculty also have access to publications, conferences, and online discussions on topics of common interest.

Develop research agendas with an international industry/academic focus: There is a dearth of research into technical communication contexts outside of North America; very little information is available on how technical writers work in other jurisdictions. While the STC magazine, *Intercom,* frequently publishes short articles on culture and international issues, we have limited information on, for example, work practices, salaries, and types of jobs available for technical writers in Europe, Asia, or Australia. Workplace research from Europe is very limited. Professional organizations for technical writers, such as STC chapters, tend to have small membership figures and wide geographical spreads, which has thus far limited the amount of data available.

Invite practitioners to speak to technical communication classes about their intercultural experience. Practicing technical writers who work on global virtual teams can make a valuable contribution to the technical writing classroom by describing the challenges and rewards of working with other cultures. Their experiences make a reality of theoretical concepts of intercultural inquiry.

DISCUSSION AND CONCLUSIONS

In this chapter, my intention has been to describe a structure that draws on the strengths of existing strategies for teaching intercultural communication, which does not isolate intercultural communication, and which merges theory and practice. The model described here is a tentative attempt to show how we have tried to merge best practices from European and North American curricula in our programs at UL. We covered established technical communication curriculum topics developed over decades in U.S. programs, and we incorporated discussion of intercultural issues into our treatment of these topics. Moreover, we moved the teaching of intercultural communication outside the classroom by enabling students and faculty to collaborate with other cultures and by seeking input from industry.

While, due to the maturity of the discipline there, North America sets the standard by which technical communication programs elsewhere are judged, the European university system benefits from the study of intercultural communication because it privileges exposure to languages and cultures other than those native to the student. Although information on university programs in technical communication outside of North America is limited, Smith (2003) describes 12 (of approximately 18) academic technical communication programs in Germany, where he considers the discipline "robust and growing." He suggests that curricula in Germany cover "the wide spectrum of documents, media, and skills that technical communicators need in the 21st century" but that they "emphasize internships, foreign languages, and study abroad programs more than programs in the United States" (p. 361). In undergraduate programs throughout Europe, and in many disciplines but especially the humanities, study and possibly work abroad are nonnegotiable requirements. Furthermore, in addition to standard content as taught in U.S. programs, European technical communication students are likely to study localization, internationalization, writing for international audiences, controlled English, and cultural awareness. Whiteside (2003, p. 309) points out that much of this content is "making [its] way" into U.S. curricula also.

In North America, a tradition of workplace research exists, which we in Europe can learn from and build on to query the European work context for technical writers. In Europe, meanwhile, language variety and cultural diversity within the EU have placed an emphasis on university structures that encourage mobility and intercultural collaboration.

Operating the structure outlined here cannot guarantee that all students will be expert communicators within all cultures in all rhetorical situations immediately upon beginning work assignments. However, students who engage with all stages of the process will almost certainly approach communication with an open mind and a better knowledge of the challenges and rewards of working internationally.

APPENDIX A: SELECTED READINGS

These readings, generally accessible, can supplement teaching on the topics listed below, adding an intercultural dimension to a traditional curriculum. As well as those readings listed here, *Intercom* frequently publishes articles on subjects related to international and intercultural communication.

Documentation Management

Hoft, N. (1995). *International technical communication*. New York: Wiley.

Ethics

Dragga, S. (1999). Ethical intercultural technical communication: Looking through the lens of Confucian ethics. *Technical Communication Quarterly, 8*(4), 365–381.

Dragga, S. (2001). A visit to the forbidden city: A sign of the times. In D. Bosley (Ed.), *Global contexts: Case studies in international technical communication* (pp. 10–18). Boston: Allyn and Bacon.

Writing Style

Flammia, M. (2005). Connecting to the audience: Strategies for teaching students to write for translation, *Proceedings of the IEEE International Professional Communication Conference,* 379–389.

Haara, B. (1998). Challenging the way we learn to write for a global audience. *Proceedings of the IEEE International Professional Communication Conference,* 293–303, 407–425.

Hoft, N. (1995). *International technical communication*. New York: Wiley.

Kirkman, J. (1992). *Good style: Writing for science and technology*. London: E. & F. N. Spon.

Rubens, P. (2001). *Science and technical writing*. New York: Routledge.

Thrush, E. A. (2001). Plain English? A study of plain English vocabulary and international audiences. *Technical Communication, 48*(3), 289–296.

Weiss, E. H. (2005). *The elements of international English style: A guide to writing correspondence, reports, technical documents, and internet pages for a global audience*. Armonk, NY: M. E. Sharpe.

Weiss, T. (1995). Translation in a borderless world. *Technical Communication Quarterly, 4*(4), 407–425.

Information, Interaction, and Graphic Design

Gribbons, W. M. (1997). Designing for the global community. *Proceedings of IEEE International Professional Communication Conference*, 261–273.

Hoft, N. (1995). *International technical communication*. New York: Wiley.

Kostelnick, C. (1995). Cultural adaptation and information design: Two contrasting views. *IEEE Transactions on Professional Communication, 38*(4), 182–196.

Tebeaux, E., & Driskill, L. (1999). Culture and the shape of rhetoric: Protocols of international document design. In C. Lovitt & D. Goswami (Eds.), *Exploring the rhetoric of international professional communication: An agenda for teachers and researchers* (pp. 305–310). New York: Baywood.

Legal Issues

World Intellectual Property Organization (WIPO). (2006). Retrieved June 4, 2008, from http://www.wipo.org

World Wide Web Consortium (W3C). (2006). *Web content accessibility guidelines (WCAG) Version 1.* Retrieved June 4, 2008, from http://www.w3.org/TR/WCAG10/

Collaboration

Davis, B. H., Brewer, J., & Chang, Y. (2001). Usage as an interactive strategy for international team building: The never-ending story. In D. Bosley (Ed.), *Global contexts: Case studies in international technical communication* (pp. 42–51). Boston: Allyn and Bacon.

St.Amant, K. (2002). When cultures and computers collide: Rethinking computer mediated communication according to international and intercultural expectations. *Journal of Business and Technical Communication, 16*(2), 196–214.

Technology

Esselink, B. (2000). *A practical guide to localization*. Amsterdam: John Benjamins.

Hoft, N. (1995). *International technical communication*. New York: Wiley.

Localisation Research Centre (LRC). (2006). Resources. Retrieved June 4, 2008, from http://www.localisation.ie/Archive.htm

Rhetoric and Culture

Barker, C. (2004). *Cultural studies: Theory and practice* (2nd ed). London: Sage.

Eagleton, T. (2000). *The idea of culture*. Oxford: Blackwell.

Hoft, N. (1995). *International technical communication*. New York: Wiley.

Sun, H. (2002). *Why cultural contexts are missing: A rhetorical critique of localization practices.* Proceedings of STC 49th Annual Conference.

Thrush, E. (2001). High context and low context cultures: How much communication is too much? In D. Bosley (Ed.), *Global contexts: Case studies in international technical communication* (pp. 27–41). Boston: Allyn and Bacon.

REFERENCES

Anderson, P. V. (2003). *Technical communication: A reader-centered approach* (5th ed.). Boston: Thomson/Heinle.

Bosley, D. (2000). *Global contexts: Case studies in international technical communication.* Boston: Allyn and Bacon.

DeVoss, D., Jasken, J., & Hayden, D. (2002). Teaching intracultural and intercultural communication: A critique and suggested method. *Journal of Business and Technical Communication, 16*(1), 69–94.

Eagleton, T. (2000). *The idea of culture*. Oxford, UK: Blackwell.

European Union. (2006). *Erasmus mundus.* Retrieved June 4, 2008, from http://ec.europa.eu/education/programmes/mundus/index_en.html

Goswami, D. (1999). Afterword: Teaching and research directions for international professional communication. In C. Lovitt & D. Goswami (Eds.), *Exploring the rhetoric of international professional communication: An agenda for teachers and researchers* (pp. 305–310). New York: Baywood.

Hall, E. T. (1977). *Beyond culture.* New York: Anchor Press/Doubleday.

Hayhoe, G. (2006). The globalization of our profession. *Technical Communication, 53*(1), 9–10.

Hofstede, G. (1994). *Culture and organizations: Software of the mind.* London: HarperCollins.

Hoft, N. (1995a). A curriculum for the research and practice of international technical communication. *Technical Communication, 42*(4), 650–652.

Hoft, N. (1995b). *International technical communication.* New York: Wiley.

Lannon, J. (2005). *Technical communication* (10th ed.). New York: Longman.

Lovitt, C. (1999). Rethinking the role of culture in international professional communication. In C. Lovitt & D. Goswami (Eds.), *Exploring the rhetoric of international professional communication: An agenda for teachers and researchers* (pp. 1–13). New York: Baywood.

Markel, M. (2003). *Technical communication.* Boston: Bedford/St Martin's.

Maylath, B., & Thrush, E. (2000). Café, thé, ou lait? Teaching technical communicators to manage translation and localization. In P. J. Hafer & H. J. Schreiber (Eds.), *Managing global communication in science and technology* (pp. 233–254). New York: Wiley.

Rabkin, B. (2005). Closing the globalization gap with DITA. *Centre for Information Development Management.* Retrieved May 29, 2006, from http://www.infomanagementcenter.com/enewsletter/200508/second.htm

Sellin, R., & Winters, E. (2005). *Cultural issues in business communication.* Berkeley, CA: Booksurge Publishing.

Smith, H. (2003). German academic programs in technical communication. *Journal of Technical Writing and Communication, 33*(4), 349–363.

Starke-Meyerring, D. (2005). Meeting the challenges of globalization: A framework for global literacies in professional communication programs. *Journal of Business and Technical Communication, 19*(4), 468–499.

Teichler, U., Gordon, J., & Maiworm, F. (2000). *Socrates 2000 evaluation study.* Paris: European Institute of Education and Social Policy. Retrieved June 4, 2008, from http://ec.europa.eu/education/programmes/evaluation/soc19.pdf

Trompenaars, F. (1993). *Riding the waves of culture: Understanding diversity in global business.* London: Nicholas Brealy.

Victor, D. (1992). *International business communication.* New York: HarperCollins.

Whiteside, A. L. (2003). The skills that technical communicators need: An investigation of technical communication graduates, managers, and curricula. *Journal of Technical Writing and Communication, 33*(4), 303–318.

CHAPTER 2

Seeing Difference: Teaching Intercultural Communication through Visual Rhetoric

Charles Kostelnick

As technical communication has become increasingly international, designing visual language for culturally diverse audiences has assumed greater importance in the field. Visual language includes many forms of print and online materials, ranging from typography, illustrations, and icons to page, screen, and data design. Although translating verbal language across cultures remains mostly the province of experts, virtually any technical communicator can translate visual language because technology has democratized design tools. By studying visual language, students can begin the process of designing for international audiences as well as learn about intercultural communication more broadly.

Instructors can accomplish these goals with their students through both analysis and design. *Analyzing* visual language across cultures heightens students' awareness of cultural differences and the rhetorical implications of how those differences manifest themselves visually through conventional practices. *Designing* visual language across cultures challenges students to adapt visual language to cross-cultural audiences and to assess the rhetorical impact of those designs with the intended readers.

OVERVIEW OF HOW CULTURE SHAPES VISUAL LANGUAGE

Teaching intercultural communication can be challenging because communication practices across cultures vary immensely. Picturing techniques, color, typography, and even public information symbols can all differ across national and cultural borders—sometimes radically and sometimes subtly, but in either case with significant rhetorical consequences for both designer and reader. Understandably, an instructor's knowledge of design variations across cultures is necessarily limited. Experiencing the "other" textually can be vicarious and bewildering because few students master a second language well enough to translate it proficiently across cultures, let alone to analyze it rhetorically. However, *visual* language can bridge that chasm, readily enabling students to *see* difference, both in print and online. An annual report of a Japanese company, a Web site for a Russian university, or even a letter from a British company can immediately enable students to distinguish, on a purely perceptual level, between what looks normal to them and what looks different, in much the same way that we notice variations in our environment (see, for example, Arnheim, 1969, pp. 20–22). In short, through the faculty of vision, *nature* hardwires us to notice difference, which can certainly supply a pedagogical springboard for teaching intercultural design.

Perception provides the portal for discovering design variations, but *nurture* provides the lens through which to scrutinize them. Through learning and experience, readers assimilate the interpretive lens of a given culture (and perhaps several cultures), and that lens profoundly shapes their vision. Defining culture is, of course, a *sine qua non* to understanding its rhetorical implications for visual design. By *culture*, do we mean ethnicity, nationality, regional or other geographical proximity, or even organizational identity? By *intercultural*, do we mean communication within cultural groups, or do we mean across two or more cultural groups, whereby differences in communication practices are acknowledged and addressed? Or by *intercultural*, do we mean communication internationally across the entire globe, whereby multiple cultures are bridged by a single communication? These issues are addressed in chapters throughout this book, and a discussion of them should precede any instruction in intercultural design. In this chapter, *culture* is mainly located in ethnic/national identity outside the United States, and *intercultural* communication spans all of the scenarios invoked above.

Teaching intercultural design transports instructors and students to novel and even exotic design domains that can be exciting visually, intellectually, and rhetorically. Still, that journey can be daunting: Trying to establish a set of design guidelines for the panoply of cultures that students are likely to communicate with in the workplace can be futile and frustrating, especially without a macrolevel framework to guide the way. As shown by Deborah Bosley (1999, pp. 267–269), one extremely useful framework for describing broad cultural characteristics and their effects on communication styles is Edward Hall's concept of "high context" and "low context" cultures (Hall, 1983, pp. 56–62). For example, in Hall's framework, members of high-context cultures prefer fewer explicit details than do members of low-context cultures like Germany (Hall, 1983, pp. 56–59). Bosley has perceptively applied Hall's high/low context spectrum to designing visual elements: To meet readers' expectations, visual representations for readers in low-context cultures should generally be more precise and explicit than those for high-context cultures (Bosley, 1999, p. 269). On the macrolevel of culture, we can also examine visual communication practices through the lenses of aesthetics (Kostelnick, 2004), ideology (Barton & Barton, 1993; Brasseur, 2003), and semiotics (Barthes, 1985; Kress & van Leeuwen, 1996).

Of course, any communicative interaction takes places on the microlevel, one reader at a time, so the broad brush strokes of culture must be sharpened with a more local and in-depth understanding of cultural context. As such, the instructor should engage students in a thoughtful *exploration* of how visual forms embody specific cultural elements and the rhetorical consequences those forms have in a given situation. For example, rather than the instructor and students analyzing or designing an instructional document for generic Western or Middle Eastern readers, they might rather do so for Dutch readers, in some depth, immersing themselves in Dutch culture, identifying conventional design forms, and perhaps even inviting online feedback from Dutch readers.

Still, knowing design conventions based on culture, however broadly or narrowly defined, cannot guarantee how designers or readers will interpret language in a given situation: Flying at 10,000 or even 1,000 feet cannot afford clear enough insight on the ground where interactions unfold, with all of their particulars and idiosyncrasies, whatever the cultural setting. As Thomas Kent (1993) has demonstrated in his theory of paralogic rhetoric, only at street level, one communicative interaction at a time, can anyone know—and imperfectly at that—how discourse participants will actually behave. This hermeneutical reality should inform and qualify the pedagogical approaches outlined below.

VISUAL ANALYSIS

Analyzing visual language across cultures can be a very powerful tool for teaching intercultural communication because it heightens students' awareness of cultural difference through the highly accessible medium of visual language (see Kostelnick & Roberts, 1998). Visual analysis can be very efficient because students can quickly immerse themselves in another culture by studying its indigenous visual language—in other words, its conventional design practices. Studying the visual language of documents that originate in a given culture is essentially studying cultural conventions: design practices that are, at least in part, learned and acquired by immersion in a given culture, which largely shapes those conventional practices (Kostelnick & Hassett, 2003, pp. 92–96). Because design conventions can become so common and taken for granted, their status as artificial constructs becomes obscured, and users "naturalize" them (Barthes, 1972, pp. 129–131; Barthes, 1985, pp. 34, 39–40; Barton & Barton, 1993).

Students can initially grasp the influence of culture on design by examining styles of ubiquitous artifacts such as buildings and cars as well as clothes that represent various periods and places, and exploring how those forms project certain cultural values. A classical Greek temple, for example, represents an aesthetic and a worldview that vary greatly from those projected by a Gothic cathedral, a Buddhist temple, or a modernist office building of steel and glass. For example, Figure 1 shows Morrill Hall on the campus of Iowa State University, a Victorian structure that illustrates the medieval revival in visual aesthetics in the 19th century. Like many public buildings of its time—libraries, hospitals, high schools, and courthouses—Morrill Hall projects the visual language of a castle, complete with a massive turret, towering gables, and a crowning cupola. Understanding the cultural origins of these forms—in Romantic literature (especially the Gothic novel), in Ruskin's social and aesthetic criticism, in the Arts and Crafts movement—can reveal the impact on design of ideas, values, and aesthetic movements tied to a particular time and place.

Other more portable and accessible artifacts can serve the same heuristic purpose. American cars and clothes from the "innocent" and early space-age 1950s, for example, represent a very different cultural mindset than do cars and clothes of the "green" and more worldly 2000s. Household furniture and appliances, wallpaper, drapes, radios and TVs—in other words, virtually anything that's designed—can serve a similar purpose. The concept of how culture shapes design can then be extrapolated to information design, which also reveals cultural characteristics that can be identified and analyzed.

At the outset of analysis, students need to recognize design elements that differ from those they are accustomed to, that look a little odd, exotic, or dissonant to them—in short, design forms that lie outside the boundaries of, or conflict with, their own visual literacy. Sometimes subtle visual elements (a degree symbol for a bullet in a list) can draw students' attention, alerting them to the distance from (or proximity to) their own visual experiences. Although students should begin by seeing the differences as carefully and precisely as possible, the goal of analysis is not to identify every visual variation so as to catalog and memorize them, much less to generate a comprehensive list of do's and don'ts. Rather the goal is to consider the observed differences as illustrative in developing an understanding of and an appreciation for how cultures differ in the way they represent information.

Like any rhetorical analysis, documents should be examined holistically—as brochures, instructional manuals, Web sites—so that their full rhetorical impact can be assessed in a specific situational context. However, focusing on a few visual elements, rather than seeing everything all at once, can result in a more systematic and effective analysis. Three particularly productive areas for intercultural analysis are pictures, text design, and color.

Figure 1. Morrill Hall, Iowa State University campus.

Pictures and Icons: Whose World is That?

Pictures are often touted as a universal language, and on a purely literal level that holds true because readers can usually identify the objects represented in pictures (see Jones & Hagen, 1980). However, pictures are typically saturated with identifiable markers that reveal the cultural orientation of their producers, such as human figures, perspective and projection techniques, environmental context, metaphors and symbols, and the proximity of pictures to text. These elements can provide productive avenues for analysis.

1. *Figures*. Human forms in pictures, if they appear, are often the single most important factor in defining cultural difference. Figures contain numerous cultural markers like skin tone, hands, dress, gestures, and gender roles, and with a little prompting, students readily identify these clues, however subtle, though they may not agree exactly on which cultures or values they represent. Figure 2, for example, shows an illustration from an instructional manual for a copy machine. The feminine hands say something about the assumptions behind the producers of this image: that women make up a large part of the workforce and that they often hold clerical jobs where they operate and are responsible for routinely maintaining copy machines—a late 20th-century gender stereotype typical in Western cultures.
2. *Perspective and other projection issues*. Drawing techniques often vary across cultures, including the spatial orientation of objects, perspective (or lack of it), and the focus on certain features of the depicted subject (Mangan, 1978). For example, instructional materials for Japanese readers

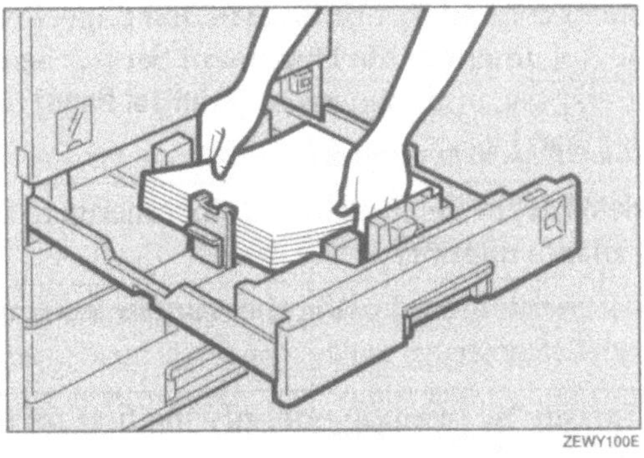

Figure 2. Illustration from an Instructional Manual for a Ricoh Aficio copy machine (p. 110). Reprinted with permission of Ricoh Americas Corporation.

often contain illustrations in cartoon style (see Fukuoka, Kojima, & Spyridakis, 1999, p. 169)—that is, without any shades or shadows or realistic elements. Even highly technical drawings can vary across cultures in their arrangement of plans and elevations of objects relative to each other (Booker, 1979, pp. 155–170, 179–181).

3. *Environmental context.* The environment in which the picture is situated provides vital clues about cultural context. A picture of a tractor in a cornfield as opposed to one in a field of sugar cane can completely change the cultural context of the reader's interpretation. On the other hand, illustrations often use line drawings with white backgrounds that erase this context, as if context were mere "noise" that diverts the reader's attention from the most important information.
4. *Visual metaphors and symbols.* Pictures, especially icons, often have metaphorical or symbolic meaning (Horton, 1993, pp. 686–688). Animals, plants, and objects of certain shapes can vary significantly in their meaning across cultures. For example, the image of a horse in the United States or western Europe can have a personal and recreational meaning, but in other cultural settings a horse may be a relatively rare or wild species or valued primarily for the functional work it does.
5. *Relationship of pictures to text.* Readers of American documents assume that a picture correlates directly with the text, clarifying or enhancing it in some way; however, this isn't always the case in, for example, Japanese documents, where the relationship may be ambiguous (Maitra & Goswami, 1995). The sheer density of the pictures on the page or screen may reveal cultural orientation. Japanese and other Eastern cultures value and expect highly picture-oriented pages (though a study by Fukuoka, Kojima, and Spyridakis, 1999, revealed that Americans have a similar preference for illustrations, at least in user manuals).

Typography: Strange Text

The typographical layout and texture of a page or screen is another area that students can assess for cultural variations in design. The inaccessibility of foreign verbal languages—such as Russian, Arabic, or Chinese—to American students can actually help students focus on the visual language of a document, though it helps if someone is available to translate verbally if necessary.

1. *Typefaces.* From the onset of printing, typefaces have always revealed some cultural, regional, or geographical trace. Today sans serif typefaces (e.g., Univers or Helvetica), a staple of modernist design, appear often in northern Europe, such as the Netherlands and France. Sometimes even a minute typographical element, like italic type or the use of a hyphen for a bullet in a list, can reveal a document's genealogy.

2. *Text layout.* Line length, spacing between lines, and the number of columns can often disclose something about the cultural derivation of a document. For example, the placement of elements on a letter—inside address, date, salutation, signature block—can signal its origin. Germans, for example, have precise conventions about formatting letters, memos, and other routine documents, and these differ from those in the United States.
3. *Shape, size, and orientation of the page.* A document printed on an A4 paper stock, which is slightly longer and more elongated than standard American paper, immediately places it outside the United States and probably in the realm of Europe, India, or Hong Kong, but certainly not the United States. Seemingly odd shapes and sizes of pages (and even how they are folded, attached, or placed in envelopes) can appear exotic and foreign to American readers.

Color: Loaded with Cultural Meaning

Color is one of the most culturally charged visual elements, infiltrating virtually every aspect of our perceptual and interpretive lives. Learning to read color is part of becoming visually literate in any culture.

1. *National or regional identity.* In America, the color scheme red, white, and blue indicates patriotism, while in Italy the color scheme of national identity is red, white, and green. Virtually every nation, as well as many regions, have comparable color associations. Oftentimes, these color schemes play a prominent role in Web design, especially in historical moments that are politically charged (e.g., the ubiquitous appearance of red, white, and blue immediately after 9/11).
2. *Religious symbolism.* Even in secular cultures, colors invoke sacred meanings among readers. For example, purple stands for the Christian Lenten season; Jews identify with blue and white, which are also the colors of the Israeli flag; and green in Islam evokes the prophet Mohammad (see Hoft, 1995, pp. 267–268). These are powerful associations with color, and lack of sensitivity to them can sometimes result in rhetorical chaos through a complete loss of clarity and ethos.
3. *A panoply of other cultural associations.* In the United States, green can stand for money (dollars), but it can also stand for the Christmas season, environmentalism, or Irish ethnicity. Black is typically associated with death and mourning in the United States, but it can also stand for many other things. For example, Mackiewiez (2007) found that Asian readers preferred PowerPoint charts with a black background more than American readers did, perhaps because of different color associations (pp. 152–153).

This brief taxonomy of visual elements can help students identify the of design differences they are likely to encounter when they interpret documents from other cultures. And conversely, recognizing those differences will also help them see that the visual conventions that they take for granted also have cultural associations.

The goal of analysis, however, is not to identify and enumerate every visual manifestation of cultural difference, which would be impossible to achieve. Instead, these observations enable students to understand and appreciate how cultures visualize information differently. Students then need to explore what those differences mean, what they reveal about the host culture, and what the rhetorical consequences are—that is, how the cultural identity invoked by visual language shapes the designer's choices and the reader's interpretation.

EXERCISES AND ASSIGNMENTS IN VISUAL/CULTURAL ANALYSIS

Instructors have a wide variety of activities with which to engage students in visual analysis, ranging from short in-class activities to full-blown assignments they can do collaboratively. All of these activities have the goal of making students more aware of intercultural issues, both design and communication more broadly, and consequently of enhancing students' rhetorical skills.

To analyze visual language, students need examples of documents from other cultures. Ideally, students should gather their own examples because they usually know more about the rhetorical situation (or other contextual factors) when they have encountered the documents themselves, even if they were not the intended audience. However, print documents that originate from outside North America may be difficult, even impossible, for them to find, especially on short notice. So the instructor might supply a set of print documents from which students can choose for their analyses or have students go online and find Web sites that originate in cultures other than their own. For example, Figure 3 shows the Web site for the University Polytehnica of Bucharest in Romania. Several visual elements—the typography, the curvilinear forms, some of the architectural elements, and perhaps even the color scheme (mostly shades of blue)—immediately differentiate it from the Web sites of American universities and provide a springboard for analyzing design based on cultural context. This Web site might also be compared with other eastern European Web sites, particularly academic ones, to see if conventional patterns emerge among them.

Visual analyses can be done in several ways, beginning with a large-group discussion of the variables listed above relative to a document chosen by the instructor. Students can also be invited to access a Web page that displays distinctive cultural traces. As they begin their analyses, students can address the following questions:

- Where in the world did the document come from? What do students know about its culture? Have any students had experience or contact with this

Figure 3. Home page for the University Polytehnica of Bucharest. Reprinted with permission of the University POLITEHNICA of Bucharest.

culture? If so, what do they know that will help them understand the visual language of this document? If not, how can they learn about this culture?

- What is the rhetorical situation for the document? Who is its audience? What is its purpose? In what situational context will it be used?
- What design elements—pictures, icons, typography, color, and such—look different, odd, or exotic about the document? How exactly do these elements differ from what looks conventional to students—that is, from what they would normally expect?

- What can students infer from these design differences about the cultural orientation of the document? What insights can the instructor share about these cultural fingerprints?
- Given the rhetorical situation, how likely will this document succeed? Is there a mismatch between the readers and their design expectations? If students or Americans are part of the intended audience, how well does the design work and how can it be improved?
- What specific recommendations can students make for designing for this cultural context?

These questions provide merely a starting point for students to explore visual language across cultures. However, as they proceed with their analyses, students should extract from them a greater awareness of how culture affects the design of even the most quotidian and functional documents, from a bill to a business card. They should also come to realize that learning about intercultural design is less the acquisition of a discrete set of facts than a process of discovery through their interactions with design artifacts and, where possible, the people who use them.

STRATEGIES FOR INTERCULTURAL DESIGN

Although visual analysis can provide a general introduction to intercultural design (and intercultural communication more generally), learning by design is a more aggressive pedagogical mode that immediately poses additional questions:

- If students are designing for a specific cultural group, what criteria will guide their design decisions?
- How can students manage to adapt a single design to audiences from a variety of cultural groups?
- How will they know how well their designs succeeded and what aspects of them need further improvement?

To provide a framework for thinking about these issues, the instructor might introduce two contemporary ways of addressing intercultural design: modern vs. postmodern (Kostelnick, 1995). In a nutshell, modernism is a one-size-fits-all approach, which seeks to find universal design forms that bridge cultural differences; and postmodernism emphasizes adaptation to specific cultural contexts. Table 1 summarizes key features of the two approaches.

Modernist Approach

Modernism was founded on the principles of universal design forms that superseded traditional conventions and erased cultural differences, which was both a political and aesthetic expediency after the cataclysmic conflict of

Table 1. Two Approaches to Intercultural Design: Modern and Postmodern

Criteria	Modernism	Postmodernism
Philosophy	Universal, one-size-fits-all	Adaptation to cultural context
Cultural orientation	Emphasis on perceptual and cognitive similarities among readers that can bridge cultures	Emphasis on cultural values and visual conventions shared by a given audience
Design elements:		
Typography	Sans serif typefaces such as Univers, Helvetica, or Avant-Garde	Typefaces that match the cultural dynamics of the situation
Human figures	High-contrast human figures without cultural markers	Cultural context invoked via clothes and environment
Page/screen design	Clean, simple design that implements Gestalt principles	Design consistent with reader's cultural expectations
Color	Black-and-white or primary colors that optimize perception	Colors that invoke the intended meanings for a given audience

World War I. These differences could partly be remedied by a design program that produced streamlined, functional forms that could be transplanted virtually anywhere in the world. In graphic design, this approach was exemplified by elements like sans serif fonts and other culturally "neutral" forms (Kinross, 1985), including simple linework with geometric forms like circles and squares and high-contrast Isotype forms, invented by Otto Neurath (1936) for displaying quantitative data. The goal of Neurath, and of modernism in general, was to appeal directly to users' innate perceptual faculties rather than learned conventions, thereby unifying users across cultures (Lupton, 1986). Gestalt principles of figure-ground contrast and grouping, for example, provided modernist designers with a perceptual and cognitive framework for implementing this approach.

In today's global economy, the modernist approach is still highly appealing because it aims toward a universal design that minimizes cultural difference by deploying forms that, in theory, can be understood by anyone and transplanted anywhere on earth. Figure 4 shows a typical modernist form: abstract, high contrast, geometrical, and even slightly mechanical. The image appears on a box containing a shelving unit that the buyer must assemble after transporting the box (with a partner) to the appropriate place. The human forms in the image are

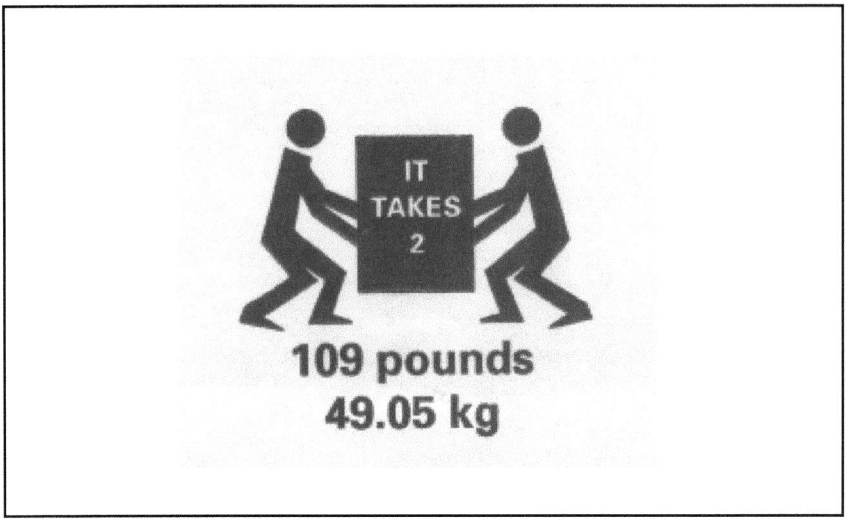

Figure 4. Modernist figures from a muscle rack heavy duty steel shelving package. Edsal Manufacturing Company, Chicago, IL, 60609.

direct and emphatic, and yet they are completely anonymous and cleansed of cultural or ethnic content.

Postmodern Approach

A postmodern approach, on the other hand, is based on the proposition that design is a communicative act with specific audiences and therefore adaptation to context is necessary for effective design (see Jencks, 2002). In terms of architectural design, for example, a building cannot just be planted in the middle of a neighborhood without the designers' attention to the visual and cultural context of that environment. Likewise, in document design, visual language must match the cultural and social context in which it is deployed. Doing so empowers readers by providing a comfortable and functional lens through which they can envision information, by acknowledging their design preferences and desires, and by addressing their ideological differences and quandaries as well (see Barton & Barton, 1993). In short, to a postmodernist, the cultural orientation of the audience matters.

Depending on the rhetorical situation, the modern and postmodern approaches have their strengths and weaknesses. The instructor can demonstrate and critique the modernist approach by showing documents that are translated verbally into several languages but that use the same visual language, within the same document or across several documents. Instructions for office equipment and consumer products provide a ready pool of examples, as do warnings and public

Figure 5. Warning label from a plastic container, Sterilite Corporation, Townsend, MA. Created by Sterilite Corporation.

information signs. Figure 5 shows an example of a warning on a plastic storage box, with the verbal language in English, French, and Spanish. The image bridges those three cultures with the high-contrast modernist form of the child and the conventional circle and slash to indicate a proscriptive message. Postmodern examples may be more difficult to find, and the Web might be the best place to seek them. Corporations, for example, often tailor their Web sites visually for a specific country or geographical or cultural region. A simple but effective technique often used on Web sites is to provide pictures of people from the host culture to bolster the site's ethos. Corporate sites based in other countries (Japan, Germany, Russia) that target an American audience might be particularly accessible and interesting to students.

So how can students decide which approach to implement in their own designs? If they have a large heterogeneous audience, and visual translations for subsets of that audience are not feasible, they might lean heavily toward the modernist approach, using forms that will appeal to a diverse audience. On the other hand, if they have a more focused and homogenous audience, they might attempt a postmodern adaptation by deploying pictures, typography, and colors that match the cultural expectations of their readers. Both approaches yield rhetorical benefits, for example, by building ethos: modernism by evoking a clean, economical, and objective look and postmodernism by enabling readers to identify with specific cultural conventions and themes.

EXERCISES AND ASSIGNMENTS IN INTERCULTURAL DESIGN

Designing interculturally can be extremely challenging, but it can greatly enhance student learning, largely because of the thinking process that it requires of students in order to orient themselves to other cultural perspectives. For design projects, instructors can select from a range of options, some of which are outlined below:

1. Design a sign for a public space that is used by a diverse international audience with a variety of cultural perspectives. For example, students could design a sign (a) for a zoo that tells visitors not to feed the animals; (b) for an art museum that tells visitors not to eat or drink; or (c) for a soccer stadium that directs spectators to the restrooms. Students could test their signs with the international students in the class or with international students in other classes.
2. Design a document for a small targeted audience from another culture, such as western European readers from, say, France. A document as simple as a one-page newsletter would require students to select paper sizes, fonts, and colors that appropriately matched their audiences. Alternately, students can design a document for an indigenous and relatively well-known audience that shares their cultural orientation and then redesign the same document for an audience from another culture.
3. Visually revise an existing document—a brochure, set of instructions, fact sheet, Web site—for an audience from a different culture. For example, if the university has a Web site for an intensive English program, how might it be designed specifically to target students from Malaysia or Indonesia? Alternately, the instructor can provide a document from another culture and ask students to redesign it for an American audience. For either approach, students can write brief self-analyses in which they describe their design decisions and the intended rhetorical impact.
4. Design a set of visual (wordless) instructions for a large international audience that includes several cultures (e.g., European, South American), particularly ones that the instructor knows something about. The instructor can provide a short set of verbal instructions that students then translate pictorially. For example, students can teach their readers how to use a piece of office equipment or a simple consumer product (or a product the instructor invents), or they can teach readers how to perform a task. This assignment can be challenging because students cannot possibly envision (nor can the instructor!) *all* of the cultural interpretations of their audiences and the manifold problems that such a mixed audience might create. This

assignment lends itself to collaboration because experiencing the design process—inventing pictures, discussing their strengths and weaknesses, revising them—supersedes achieving the perfect solution. Ideally, each group would include an international student or a student with international experience. Students can also benefit greatly by sharing their results with each other and explaining their design decisions.

One of the most challenging aspects of these assignments for the instructor is assessing how well students' designs will actually work. And this can often befuddle an instructor, who may try to role-play the cultural orientation of the intended audience. One way to avoid this dilemma is to test the students' designs with readers who actually share the cultural values of the intended audience. In this regard, international students can be a tremendous resource in the class by playing the role of "expert" readers. Another valuable assessment tool is to require students to compose brief rhetorical analyses of their design processes in which they provide rationales for their design decisions. Particularly in this kind of assignment, articulating the decisions behind the design can externalize the level of the students' intercultural thinking, as well as stimulate class discussion and a fruitful dialogue with the instructor.

CONCLUSION

Teaching intercultural design is critical today because visual design encompasses so many forms and media, including the Internet, and because it extends well beyond the cultural domain of any individual. Because the visual is so ubiquitous and accessible, instruction in intercultural design can also open a window to broader issues in cross-cultural communication. That instruction can be stimulating, both visually and intellectually, but it can also be daunting and uncomfortable because most instructors command only a limited pool of knowledge of the subject.

As a result, instructors should set about *exploring* intercultural design *with* their students, foregoing any illusions about establishing a set of design guidelines for the array of cultures students are likely to encounter in the workplace. Rather, the instructor should engage students in thoughtful consideration of how visual forms embody cultural elements, how those elements shape the rhetorical process, and in turn, how students might apply that knowledge to their own designs. Enabling students to visualize cultural difference, moreover, might have profound consequences, both intellectually and practically, if it broadens their worldviews.

REFERENCES

Arnheim, R. (1969). *Visual thinking*. Berkeley: University of California Press.

Barthes, R. (1964/1985). Rhetoric of the image. *The responsibility of forms: Critical essays on music, art, and representation* ®. Howard, Trans., pp. 21–40). New York: Hill and Wang.

Barthes, R. (1972). *Mythologies* (A. Lavers, Trans.). New York: Hill and Wang.

Barton, B. F., & Barton, M. (1993). Ideology and the map: Toward a postmodern visual design practice. In N. R. Blyler & C. Thralls (Eds.), *Professional communication: The social perspective* (pp. 49–78). Newbury Park, CA: Sage.

Booker, P. J. (1963/1979). *A history of engineering drawing*. London: Northgate.

Bosley, D. S. (1999). Visual elements in cross-cultural technical communication: Recognition and comprehension as a function of cultural conventions. In C. R. Lovitt & D. Goswami (Eds.), *Exploring the rhetoric of international professional communication: An agenda for teachers and researchers* (pp. 253–276). Amityville, NY: Baywood.

Brasseur, L. E. (2003). *Visualizing technical information: A cultural critique*. Amityville, NY: Baywood.

Fukuoka, W., Kojima, Y., & Spyridakis, J. H. (1999). Illustrations in user manuals: Preference and effectiveness with Japanese and American readers. *Technical Communication, 46,* 167–176.

Hall, E. T. (1983). *The dance of life: The other dimension of time.* Garden City, NY: Anchor Press/Doubleday.

Hoft, N. L. (1995). *International technical communication: How to export information about high technology*. New York: John Wiley & Sons.

Horton, W. (1993). The almost universal language: Graphics for international documents. *Technical Communication, 40,* 682–693.

Jencks, C. (2002). *The new paradigm in architecture: The language of post-modernism.* New Haven, CT: Yale University Press.

Jones, R. K., & Hagen, M. A. (1980). A perspective on cross-cultural picture perception. In M. A. Hagen (Ed.), *The perception of pictures: Vol. 2. Dürer's devices: Beyond the projective model of pictures* (pp. 193–226). New York: Academic Press.

Kent, T. (1993). *Paralogic rhetoric: A theory of communicative interaction*. Lewisburg, PA: Bucknell University Press.

Kinross, R. (1985). The rhetoric of neutrality. *Design Issues, 2*(2), 18–30.

Kostelnick, C. (1995). Cultural adaptation and information design: Two contrasting views. *IEEE Transactions on Professional Communication, 38,* 182–196.

Kostelnick, C. (2004). Melting-pot ideology, modernist aesthetics, and the emergence of graphical conventions: The statistical atlases of the United States, 1874–1925. In C. A. Hill & M. Helmers (Eds.), *Defining visual rhetorics* (pp. 214–242). Mahwah, NJ: Erlbaum.

Kostelnick, C., & Hassett, M. (2003). *Shaping information: The rhetoric of visual conventions.* Carbondale, IL: Southern Illinois University Press.

Kostelnick, C., & Roberts, D. (1998). *Designing visual language: Strategies for professional communicators*. Needham Hts., MA: Allyn and Bacon.

Kress, G. R., & van Leeuwen, T. (1996). *Reading images: The grammar of visual design.* New York: Routledge.

Lupton, E. (1986). Reading Isotype, *Design Issues, 3*(2), 47–58.

Mackiewicz, J. (2007). Perceptions of clarity and attractiveness in PowerPoint graph slides. *Technical Communication, 54,* 145–156.

Maitra, K., & Goswami, D. (1995). Responses of American readers to visual aspects of a mid-sized Japanese company's annual report: A case study. *IEEE Transactions on Professional Communication, 38,* 197–203.

Mangan, J. (1978). Cultural conventions of pictorial representation: Iconic literacy and education. *Educational Communication and Technology—A Journal of Theory, Research, and Development, 26,* 245–267.

Neurath, O. (1936). *International picture language: The first rules of Isotype*. London: Kegan Paul, Trench, Trubner.

Ricoh Corporation. (2003). *Operating instructions/General settings guide, Aficio™ 2022/2027.*

University Polytehnica of Bucharest. Retrieved June 2008 from University Polytehnica of Bucharest Web Site: http://www.pub.ro/

CHAPTER 3

Technical Communication in India: Through the Lens of Intercultural Rhetoric

Poornima Padmanabhan

In their 2002 essay, "Cultural Studies: An Orientation for Research in Professional Communication," Charlotte Thralls and Nancy Blyler employ developments in feminist and poststructuralist studies to make a case for a broader view of cultural studies. At the core of their argument is the fact that every cultural representation is formed by a series of linkages with other discourses (social, cultural, historical, technical, and so on), and is therefore not an isolated entity. Rather, there is a larger ensemble contingent upon the several (usually complex) relationships that form the whole (Thralls & Blyler, 2002). This notion of interlinked discourses can help researchers in intercultural communication better understand the various factors that affect technical communication practices in India. In this chapter, I use Thralls and Blyler's ideas as a framework for providing readers with a brief background on some factors that significantly influence intercultural communication practices in India. In so doing, I also highlight some rhetorical practices that Indian technical communicators have developed to work in intercultural project setups. To achieve these objectives, I draw upon my own experiences working in Indian technical communication teams. In addressing these objectives, I will propose a set of strategies that academics and practitioners in technical communication can further develop to create seamless models for intercultural collaboration, communication, and education that can be applied in a variety of international business and educational settings.

BACKGROUND TO THE FIELD OF TECHNICAL COMMUNICATION IN INDIA

The field of technical communication is witnessing a rapid surge of growth in India and this growth is driven by two main factors:

1. The prominence of Indian technology-solutions firms, which preempt associated technical and end-user documentation solutions.
2. The refinement of outsourcing models, which enable companies to make business decisions about outsourcing technical communication solutions.

While no formal scholarly historiography of technical communication in India is currently available, delving into experiential accounts of the evolution of technical writing in India provides valuable insights into the ways in which the profession is practiced.

Technical communication as a profession in India is relatively young, having achieved prominence in the 1990s as a supporting function within information technology firms (Bhatia, 2005). In the year 2000, 95% of technical communicators worked in the software industry (Kamath, 2000). Technical writing in other manifestations certainly did exist in India long before the IT phenomenon in the 1990s, largely within governmental, manufacturing, and regulated industries. In 2003, there were an estimated 2,500 to 3,000 technical communicators in India (Biswas, 2003). Estimates today suggest that number has more than doubled (Chattopadhyay, 2007). Studying this rapid growth is important, since it reveals the rhetoric behind intercultural communication in India and puts both successful and failed intercultural project experiences in the right perspective.

Extending the Thralls and Blyler (2002) idea of interlinked discourse to encompass rhetoric in addition to cultural studies reveals that technical communication in India is bound by even more dependencies (linkages) on social behavioral expectations, perceptions of ethics, educational foundations, and organizational cultures. In various studies, scholars have called for models of rhetoric and cultural studies in technical communication to be more inclusive of one factor or the other, and warned of the potential for cultural studies to be reductionistic. Oversimplification, essentialism, and ethnocentrism are easy traps to fall into when discussing culture (Hunsinger, 2006). Some argue that, in order to reveal the ideological root of a culture's communicative practices, it is necessary to illuminate the ethics of different civilizations (Dragga, 1999). Others call for an "economistic version of cultural studies" to account for sociocultural, economic, and political trends (Longaker, 2006). Clearly, a unified model of culture based on this idea of linkages is yet to be devised. For this reason, this chapter cannot cover all the possible linkages/dependencies that affect technical communication practices in India. Rather, I have limited the scope of this chapter to discuss three central factors that influence how technical communication is practiced in India today. Those central factors are education, the organization, and the economy.

Education

India's technical communicators are multilingual individuals who are educated in English, but who are also often proficient in one or more Indian regional language (Kamath, 2000). Early English education in India began in the 17th century when the British reign brought with it British missionaries (Baldridge, 2002). By the 20th century, English became the official academic language in many primary and higher education institutions in India, and today it is considered an official language of the Union of India, along with Hindi (The Official Languages Act, 1963). In 1994, approximately 35 million people in India (4% of the total population) used English for daily communication (Hohenthal, 2003). While this number might appear small in relation to the nation's overall population of over 1 billion, the fact that this number accounts for tens of millions of persons makes it highly significant on an international scale where English is the language of business and technology. These Indian English speakers, moreover, attend institutions (primary, secondary, and college-level) that administer their curricula completely in English. Technical communicators are among this small subset of Indians who have educational backgrounds in English. Additionally, these Indian technical writers have undergraduate or graduate degrees in engineering, science, English, journalism, the social sciences, or the humanities.

Unfortunately, India lacks formal undergraduate and graduate programs in technical communication. A 2005 Salary Survey conducted by the Indian chapter of the Society for Technical Communication (STC) revealed that, in India, six times as many respondents did not have an educational background in technical communication than the ones who did (STC India, 2005). In the absence of formal education, individuals and for-profit institutions have set up certificate and training programs to provide needed instruction to the growing number of technical communicators being hired by Indian technology industries. While some of these programs are reputable, others are of a questionable nature, as their focus is on tool-based training vs. a mix of theory and skills. Although programs in user interface design, visual design, new media, information design, and human-computer interaction (HCI) do exist in India as a result of such industry demand, there is a surprising absence of programs dedicated to technical communication (Gummaraju, 2005). Only two institutions in India offer university-level courses in technical communication, and neither university offers an overall degree in technical communication (STC India, 2006).

Interestingly, research in intercultural technical communication education reveals that a growing number of international employers are interested in, and accepting of, online degree programs in technical areas such as technical communication (St.Amant, 2006). Moreover, a recent feasibility survey indicates that many Indians have a positive reaction to the prospects of participating in online technical communication programs from countries that already have strong technical communication programs (Baake & Wang, 2006). The possibilities for technical communication education in India are many, and practitioners recognize the need for haste in providing formal education in the field. At this juncture, the status of technical communication education raises the crucial question: how are professionals in India trained in technical communication? The answer lies with the organizations that recruit the writers.

The Organization

Technical communication groups in Indian organizations are faced with the task of not only providing on-the-job training to keep technical communicators' skills current, but also carrying out introductory training (particularly for entry-level writers). This educational situation has been an investment that technology firms have willingly made, and one in which the organization has assumed a bigger role than educational institutions in shaping the technical communication competencies of writers in India. Within this context, a major question becomes, how can organizations offer focused on-the-job training to provide writers with not only skills-based technical communication training but also with the added rhetorical competencies needed to perform theoretical studies on complex assignments involving intercultural understanding?

In reality, this is the core challenge for the field of technical communication in India. The way Indian organizations currently offer such training is to use a practical case-based project model to provide writers with a hands-on grasp of the writing life cycle and process, quality considerations, intellectual property issues, project planning methods, communication best-practices, and customer (and user) interaction. Within such a system, organizational training programs in India can last for months and run in parallel to a writer's project-based experience.

Mentoring by senior writers is an important part of organizational learning in Indian teams, and many writers begin their careers working in team setups before moving on to handle projects independently. However, demographics from the 2003 STC India Salary Survey indicate that only about 5% of technical communication professionals in India have more than 15 years experience in the field (Kamath, 2003). Thus, the ratio of potential mentors to writers in need of mentoring is low. Statistics like these draw attention to how young the profession really is in the country, but they are also symbolic of an important developmental milestone in the history of Indian technical communication.

Looking back at the history of technical communication, the current scenario in India mirrors the growth of technical communication in the United States before the advent of formal education programs in the field. In the 1950s, significant efforts went into launching and formalizing U.S. technical communication education programs, such as the establishment of dedicated technical communication programs at Rensselaer Polytechnic, MIT, and the University of Michigan (Connors, 1982). When I compare technical communication education in India today with the United States in the 1950s, it is to illustrate that the field in India today is poised at the same historical juncture, metaphorically speaking, in that this represents a time when immense thought needs to be put into building technical communication theory and education in India. The organization does have an important role in grooming technical communicators in India, but an educational foundation will simplify the organization's task and furnish writers who are better equipped to work in intercultural teams.

The Economy

Outsourcing in technological fields has become an established work model in India, especially so in information technology (Workman, 2006). Clearly, the trickle-down effect of larger macroeconomic trends in information technology, in particular, the outsourcing movement, into the field of technical communication has implications for the way Indian technical communicators view their daily work. The growing diversity of workforces and users also plays an important role in shaping organizational communicative practices. Like their peers in other parts of the world, Indian technical communicators are all too often in assignments where their target user groups are international and interdisciplinary. Many times, the project teams in which these Indian employees work are themselves international, with technical communicators from one or more companies and nations working together on the same project.

In this climate of diversity, the most effective technical communication solutions are the ones rooted in rhetorical awareness and intercultural appreciation. When writers are a part of interdisciplinary groups (like user experience groups) within larger companies, they benefit from the existing literature and case-based knowledge on user research and usability testing that they get from HCI practitioners and usability researchers.

Since intercultural interaction is a given in the current economy, the most successful Indian technical communication teams (like their counterparts in other nations) will be cognizant of and adopt user research practices that originate from the social sciences such as education, psychology, and anthropology. Until formal educational programs are available, Indian companies should equip themselves with the ability to provide this well-rounded intercultural appreciation within user-centered professions. Similarly, overseas educators teaching their students to interact effectively with Indian colleagues could draw important lessons from cultural anthropology and case-based learning accounts provided by Indian technical communication teams.

RHETORICAL PRACTICES OF INDIAN TECHNICAL COMMUNICATORS

Studying the ways in which technical communicators worldwide perceive the role of rhetoric in their work can answer important questions about culture and cultural identity. Indeed, technical communicators often participate in international teams composed of individuals from diametrically different rhetorical/communication traditions. Within such contexts, technical communicators often develop communication artifacts (e.g., documentation) for users from other rhetorical traditions. For these reasons, it is worth asking how these communicators step out of their own rhetorical and discourse communities to create successful communication solutions in the global economy. To take that line of questioning further, how do internationally distributed technical communication teams, each with their own rhetorical and organizational cultures, parse their collective intercultural knowledge?

Scholars have found anthropologist Edward T. Hall's context model of communication particularly useful when conceptualizing intercultural communication. The context model has been considered a valuable starting point in understanding behavior and cultural identities in order to develop cross-cultural understanding (St.Amant, 2001). For instance, India is classified as high-context and highly collectivist, while the United States is low-context and individualistic. Low-context cultures have been found to employ direct, explicit models of communication, whereas high-context cultures tend to employ implicit models of communication based on an assumed shared collective understanding. Context models reveal to some extent the motives behind directness or indirectness, explicit or implicit communication, persuasion, and negotiation (Jokinen & Wilcock, 2006). Cross-cultural studies in advertising strategies have found that these factors do result in differences in the communication approaches used in India and the United States. In one study, for example, it was found that while Indian advertisements used an indirect rhetorical approach and collective visual portrayals, U.S. advertisements used direct rhetorical styles based on individualistic portrayals (Niaz, 1996). Despite such findings, researchers in intercultural communication also suggest using the context model as *one* available tool and not as the authoritative standard in studying intercultural communication. Such a perspective helps to avoid the tendency to form cultural stereotypes.

While it would be correct to say that traditionally, Indian verbal and nonverbal rhetorical patterns can be classified as indirect, it must also be considered that Indian modern culture has assimilated (and continues to assimilate) influences from Western media, business models, and communication styles. The result is an intermingled rhetorical style that is much more complex to classify. The nature of outsourcing itself calls for professionals to adapt to new and cross-cultural rhetorical styles, which is especially true of technical communicators in India, since individuals usually work in intercultural collaborations.

As a result of these factors, there is no one answer to encapsulate the way Indian technical communicators view intercultural rhetoric. Many times, intercultural rhetorical strategies are intuitive heuristic-based responses to intercultural scenarios. For example, an Indian technical communicator creating user assistance for a Japanese audience would use an audience analysis heuristic to collect and analyze audience data (demographic, task based, and context based). The writer would use the heuristic to make design, branding, and content choices for the audience, but would also make some intuitive decisions about the style and tone to use (based on some level of literature study and knowledge of internationalization or localization).

Although the heuristic-based approach does derive from qualitative and quantitative user research, it has previously led to oversimplified perceptions of intercultural rhetoric. Hunsinger (2006), for example, argues that the cultural heuristics approach used to define and interpret cultures needs to be supplemented with anthropological and sociological perspectives in cultural studies. Studying anthropological and sociological perspectives in technical communication mandates some form of formal education in technical communication, which is one more way that Indian writers will benefit from the establishment of educational programs. Even so, Indian technical communicators do have a toolkit of successful intercultural rhetorical practices. It is, however, also true that there are other practices that have negatively impacted intercultural projects involving global teams.

Table 1 and Table 2 highlight some successful and problematic intercultural rhetorical practices of Indian technical communication teams.

Table 1 lists some intercultural rhetorical practices that have been used successfully by Indian technical communication teams.

Table 2 lists some intercultural rhetorical practices that have met with either negative responses or results. Some of these practices highlight problem areas for even leading technical communication groups, although they represent the biggest problem in firms that are still developing hiring and training policies or those that compromise on the quality of the writers they recruit.

Most of the practices noted in Table 1 and Table 2 are heuristic-based when employed by Indian writers and do not stem from theoretical studies. A number of these practices, however, do implicitly exhibit rhetorical considerations. For this reason, I have stated the underlying rhetorical consideration that seems to contribute to the strategy's success or failure. It would be worthwhile to conduct both quantitative investigations and qualitative explorations to better understand the relevance and impact of the practices listed in these two tables, as well as their effect on return on investment (ROI).

Table 1. Successful Intercultural Rhetorical Practices

Rhetorical practice	Reason for success
Bringing the user into the development life cycle by incorporating user research, user testing, and usability testing phases into the life cycle.	Implies that meaning is a two-way negotiation rather than a transmission from author to audience, in keeping with the articulation model of authorship (Slack, Miller, & Doak, 1993). Discourse can be viewed as being developed through a series of interactions. For instance, writers gather iterative user feedback through acceptance testing, customer service call logs, and online surveys.
Classification of users into user types based on role, expertise, and discourse conventions.	Shows that technical communicators consciously use different rhetorical conventions for users and customers and customize communication based on their specific needs.
Selection of communication media based on user and task analysis. For example, choice of printed versus online documentation or wikis vs. Web sites.	Depicts a user-centric awareness in making information design decisions.
Use of internationalization and localization standards from literature and case studies.	Depicts that awareness of the discourse in international technical communication forums exists, and best practices are being studied and customized for projects implemented.
Customization of broader organizational style guide (and industry style guides) at the start of each project.	Indicates that writers are (subconsciously perhaps) aware of the role of the style guide as a mediational artifact and use it as a tool of negotiation and clarification in their daily work.
Visualization of projects in terms of business scenarios rather than tasks and the use of case studies in documentation.	Shows that writers are aware of trends in conceptualizing technical communication as well as shifts in thinking (for example, the shifts from interface-based to task-based to scenario-based documentation).
Use of comparative models as a way to distinguish language and stylistic standards across cultures or countries.	Reflects a basic inclination toward comparative and contrastive rhetoric. This knowledge is currently very simplistic but shows potential for development with formal education.

Table 1. (Cont'd.)

Rhetorical practice	Reason for success
Use of distributed teams to situate writers and the user in the same location.	Shows that technical communicators are aware of the need for writers to interact closely with users, especially in intercultural project setups.
Illustrating activities, actors, and tasks diagrammatically; using diagrams to represent actors, activities, tasks, communication zones, life cycle, interaction points, and timelines.	Reflects the ability to conceptualize and clarify complex relationships and interactions in project life cycles.
Using experiential activities (such as project-learning and mock projects) to equip technical communicators with competencies to work in intercultural projects.	Reveals attempts to provide a realistic context of daily work in multicultural settings, although studies point out that without clear objectives or formal assessment, the value of such activities can be undermined (Matveeva, 2006).

STRATEGIES FOR INTERCULTURAL COLLABORATION

Building an appreciation for intercultural rhetoric in organizational and academic settings presents a challenge for practitioners and academics globally. An intercultural perspective devoid of stereotyping or oversimplification requires an understanding of theoretical discourses in diverse fields, including sociology, applied linguistics, rhetoric, and philosophy (Matveeva, 2006). A recurring premise across research exploring ways to cultivate intercultural appreciation is to provide an understanding of the social contexts of culture to create realistic expectations of behavior in communicative settings (Jokinen & Wilcock, 2006). A grounding in the theoretical foundations of technical communication coupled with experiential learning exercises could provide technical communication students and practitioners (both in India and in other nations) with a toolkit with which to work better in intercultural setups. Drawing from intercultural theory and case-based learning, I propose the use of seven intercultural strategies to promote interactions and exchanges with Indian technical communication teams. These strategies can be used individually or in combination by Indian and international technical communication teams and can facilitate effective exchanges across such teams. Researchers studying technical communication

Table 2. Potential Intercultural Rhetorical Problem Areas for Indian Technical Communicators

Rhetorical practice	Reason for failure
Scholarly work in theory is not used to form models of rhetorical analysis.	Prevents new ways of using theoretical models to perceive communication and daily tasks on projects and in turn, encourages an oversimplistic understanding of the field.
Approach to user research and usability testing is heuristics based.	The approach provides metrics but does not take into account anthropological or sociological studies that can provide writers with fresh perspectives on intercultural rhetoric.
Intercultural rhetorical awareness is trial and error, and hence, there is a hit-and-miss failure or success rate on projects.	Lack of a formal holistic rhetorical model to conceptualize verbal and written communication, negotiation, and intercultural project setups. Thus, ancillary but important intercultural skills (such as negotiation skills) are not theorized and analyzed.
Intercultural communication guidelines are simplistic in many cases, taking a do's and don'ts approach to understanding culture.	Oversimplification in intercultural studies is a known danger highlighted in scholarly work (Hunsinger, 2006).
Some organizations "blind-allocate" technical communicators to projects to meet demands without assessing the business risks associated with the lack of intercultural knowledge.	Lack of importance to intercultural communication reflects as lack of commitment or competence on the part of the organization.

in other cultures can adopt these strategies as well, assess their efficacy, and customize them as needed. The strategies themselves will stay the same irrespective of the culture being studied; what will change is the way they can be implemented across cultures.

Strategy 1: Arrange experiential learning activities in intercultural teams for students and technical communicators.

Matveeva (2006) explores themes for intercultural training to span topics such as cultural conventions, stereotyping, interpersonal communication, as well as organizational written and verbal communication. It would benefit technical communicators to work together on repeated assignments and experience firsthand the similarities and differences in approaches to audience, writing process, style, and organizational communication. St.Amant (2003) outlines practical strategies that communicators can use when working across cultures in international virtual offices (IVOs). Exercises in actual or simulated IVOs would provide technical communicators a chance to collaborate on real-world problems and reach mutually acceptable outcomes. Academics and organizations should consider international exchange programs for students and practitioners to countries such as India, where the profession is in a developing stage. Where travel is not possible, case-based learning structured on actual scenarios can be used depending on the context.

Strategy 2: Use comparative and contrastive rhetoric to acquaint individuals with culture-specific discourse.

Jansen (2002) wrote of the three main methods used to study the role of culture in discourse, of which, comparative studies of discourses is the first. Scholars in technical communication, linguistics, and rhetoric have all proposed the use of comparative and contrastive ways of studying rhetorical practices. The ideal way to implement this strategy is for technical communication practitioners to document their own cultural practices and expectations in workplace and interpersonal settings, and then contrast the results with similar observations carried out by teams from other cultures. Collaboration is crucial in an exercise of this nature; if done with a degree of thoroughness, comparative research can yield the kind of qualitative results technical communicators need to develop ways of working outside their own cultural circles. Another way to study culture-specific discourse is to keep track of the discourse in technical communication in local technical communication societies and publications; for example, activities and publications of the Society for Technical Communication (STC) India chapter.

Strategy 3: Work with intercultural teams to create and critique discourse in intercultural rhetoric.

It is important to work with intercultural teams to create discourse in intercultural rhetoric. Just as two individuals from the same culture can bring different but valuable viewpoints to a problem, so it is with collaborators across cultures. Creating standards, guidelines, and theoretical models in intercultural groups can reveal assumptions or stereotypes that have been tacitly accepted, and provide ways to create an inclusive model to conceptualize intercultural rhetoric. Since user research and usability testing are now a part of most technical communication life cycles, it would also be worthwhile to observe a cycle of user studies carried out by another cultural group to gauge their perceptions of intercultural user research, and then collaboratively discuss and refine intercultural user research models. This approach would help improve the standard of intercultural user-assistance solutions.

Strategy 4: Use workplace ethnographic studies to develop an understanding of communicative practices and behavior in India.

Ethnographic studies (within specific researcher and participant conditions) have been considered a favorable approach to studying human communication in intercultural settings (Crawford, 1987). Conducting and sharing ethnographic studies on technical communication work practices across cultures can help set accurate expectations of cultural communication models. More importantly, the findings from such studies can explicate not only the task workflow, but the several exchanges (verbal, written, gestural) that typify communicative events. As participant observers or even by studying ethnographies conducted by others, technical communicators can develop realistic mental models of minute-to-minute communication exchanges, rhetorical practices such as negotiation and mediation, and organizational workflows. Along with ethnographic studies, organizational theories can be used to devise frameworks of intercultural communication that can be adopted by technical communicators in work situations (Constantinides, St.Amant, & Kampf, 2001).

Strategy 5: Make the discourse and best practices in intercultural rhetoric readily available to practitioners from other cultures.

Subbiah (1997) and scholars he cites argue that the development of intertextual connections comes from an understanding of economic, historical, and sociological factors. For the establishment of any intercultural intertextual discourse, technical communication academics and practitioners must strive to make their individual discourses available to other cultures, and simultaneously provide critical feedback on their discourses. Publishing journal papers about experiences with Indian communicative practices would be of immense value to writers across the world, and Indian practitioners can significantly contribute to the ongoing discourse.

Informal publishing on the World Wide Web is another potential way to collaborate and share experiences. The evolution of Web technologies that support participation and sharing on the Internet, often collectively labeled as Web 2.0 (O'Reilly, 2005), has offered greater access to more informal sources of interaction such as blogs, wikis, and discussion forums. While many of these forums can be very useful, care should be taken to evaluate the merit of content to be sure that claims about technical communication work in other countries have some sort of empirical or experiential support.

Strategy 6: Introduce the rhetorical practices of Indian technical communicators into international curricula.

I highlighted some rhetorical communicative practices of Indian technical communicators in this chapter. However, there are numerous other rhetorical practices whose study will benefit the field at large. Academics and practitioners can start a process of intercultural acculturation in their own countries by studying these rhetorical practices further and providing the lessons learned to students and practitioners, either by way of revised curricula or organizational training.

Strategy 7: Collaborate to set up intercultural educational opportunities and approaches.

Collaboration between technical communication academics across cultures will certainly help keep curricula current and relevant. In the absence of technical communication education in India, it would still be a valuable exercise for academics (and their students) to collaborate with Indian technical communicators to exchange perspectives on outsourced technical communication models and how these affect daily project practices. Simultaneously, students expressing an interest in international experiences can consider internships with Indian technical communication firms to develop a two-way dialogue in intercultural studies. Academics exploring partnerships to establish technical communication educational programs in India (including online programs) would find it most beneficial to initiate a dialogue at this point in time and thus make valuable contributions to the profession in a new culture.

CONCLUSION

The strategies proposed in this chapter are based on my own experiences working in India and my observations of the current state of technical communication discourse in India. Situated in a business climate rooted in cross-cultural partnerships, and without any formal education to assist them, Indian technical communicators have had to develop highly job-specific strategies to execute projects successfully. There have been both heartening success stories and unfortunate misunderstandings as a result of their rhetorical practices. In either case, sweeping generalizations need to be replaced by purposeful studies that will help practitioners evaluate potential collaborations and help academics foster the discourse on intercultural rhetoric.

Academics can also evaluate whether incorporating these international intercultural rhetorical practices into curricula will benefit students who will work in largely intercultural contexts. Understanding technical communication approaches across the world using ethnographic and anthropological studies has never been as important than it is in these times of intercultural collaborations; and studying the rhetoric that underpins intercultural communication will help both practitioners and academics develop more informed strategies for practice and instruction.

REFERENCES

Baake, K. R., & Wang, J. (2006). Offshoring and serving the needs of Indian technical communication programs: A feasibility study. *Technical Communication, 53*(4), 427–438.

Baldridge, J. (2002, June–July). Linguistic and social characteristics of Indian English. *Language in India, 2*. Retrieved June 29, 2008, from http://www.languageinindia.com/junjul2002/baldridgeindianenglish.html

Bhatia, N. (2005). Emerging from the sidelines: The Indian technical writing community in a global marketplace. *INDUS (Newsletter of the STC India Chapter), 7*(4). Retrieved June 29, 2008, from http://www.stc-india.org/indus/072005/neeraj_article.htm

Biswas, N. (2003). *The future of technical writing in India.* Paper presented at the STC India 5th annual conference, Pune, India. Retrieved June 29, 2008, from http://www.stc-india.org/conferences/2003/papers/Future.ppt

Chattopadhyay, S. (2007, January 18). Simplifying jargon. *The Telegraph (India).* Retrieved June 29, 2008, from http://www.telegraphindia.com/1070118/asp/careergraph/story_7274482.asp

Connors R. J. (1982). The rise of technical writing instruction in America. In J. Johnson-Eilola & S. A. Selber (Eds.), *Central works in technical communication* (pp. 3–19). New York: Oxford.

Constantinides, H., St.Amant, K., & Kampf, C. (2001). Organizational and intercultural communication: An annotated bibliography. *Technical Communication Quarterly, 10,* 31-58.

Crawford, L. (1987, May 18–21). *Intercultural communication study: A description of an interpretive/ethnographic approach.* Paper presented at the 78th annual meeting of the Eastern Communication Association, Syracuse, NY.

Dragga, S. (1999). Ethical intercultural technical communication: Looking through the lens of Confucian ethics (electronic version). *Technical Communication Quarterly, 8*(4), 365–381.

Gummaraju, A. (2005). Information architecture concepts for the technical writer. *INDUS (Newsletter of the STC India Chapter), 7*(3). Retrieved June 29, 2008, from http://www.stc-india.org/indus/062005/themeanupama.htm

Hohenthal, A. (2003, May). English in India: Loyalty and attitudes. *Language in India, 3.* Retrieved June 29, 2008, from http://www.languageinindia.com/may2003/annika.html

Hunsinger, P. R. (2006). Culture and cultural identity in intercultural technical communication. *Technical Communication Quarterly, 15*(1), 31–48.

Jansen, D. E. (2002). Intercultural communication in engineering: A research programme to investigate the cultural influences in the negotiation of engineering projects. *World Transactions on Engineering and Technology Education, 1*(1). Retrieved June 29, 2008, from http://www.eng.monash.edu.au/uicee/worldtransactions/WorldTransAbstractsVol1No1/Microsoft%20Word%20-%2003_Jansen.pdf

Jokinen, K., & Wilcock, G. (2006). Contextual inferences in intercultural communication. In M. Suominen et al. (Eds.), *A man of measure: Festschrift in honour of Fred Karlsson* (pp. 291–300). Retrieved June 29, 2008, from http://www.ling.helsinki.fi/sky/julkaisut/SKY2006_1/1.4.4.%20JOKINEN%20&%20WILCOCK.pdf

Kamath, G. (2000, May). The India paradox. *Intercom,* 10–11.

Kamath, G. (2003, September 1). Technical writers' salary survey 2003. *IT People.* Retrieved June 29, 2008, from http://www.expressitpeople.com/20030901/careers1.shtml

Longaker, M. G., (2006). Back to basics: An apology for economism in technical writing scholarship. *Technical Communication Quarterly, 15*(1), 9–29.

Matveeva, N. (2006). *Teaching intercultural communication in a service technical writing course: Alternative ways of presenting intercultural issues in technical writing textbooks and in real classrooms.* Retrieved June 29, 2008, from Texas Tech University, Electronic Theses and Dissertations Web site: http://etd.lib.ttu.edu/theses/available/etd-11062006-172848/unrestricted/MatveevaDissertationDec6.pdf

Niaz, A. (1996). *Cross-cultural content analysis of advertising from the United States and India.* Retrieved June 29, 2008, from http://www.dissertation.com/book.php?method=ISBN&book=1581120842 (ISBN 1581120842).

The Official Languages Act. (1963). The Official Languages Act, 1963 (as amended, 1967), Act No. 19 of 1963. *Language in India, 2*(2). Retrieved November 18, 2009 from http://www.languageinindia.com/april2002/officiallanguagesact.html

O'Reilly, T. (2005). *What is Web 2.0: Design patterns and business models for the next generation of software.* Retrieved June 29, 2008, from http://www.oreillynet.com/pub/a/oreilly/tim/news/2005/09/30/what-is-web-20.html

Slack, J., Miller, D., & Doak, J. (1993). The technical communicator as author: Meaning, power, authority. *Journal of Business and Technical Communication, 7*(1), 1–36.

Society for Technical Communication (2005). *2005: Salary survey of Indian technical communicators.* Retrieved June 29, 2008, from http://www.stc-india.org/activities/surveys/STC_India_2005_Salary_Survey.pdf

Society for Technical Communication (2006). [Home Page]. Retrieved February 23, 2007, from http://stc-india.org/

St.Amant, K. R. (2001, April). Identity and international online communication. *Intercom,* 16–17.

St.Amant, K. R. (2003, April). Communication in International Virtual Offices. *Intercom,* 27–28.

St.Amant, K. (2006, May). Internationalizing online training. *Intercom,* 18–21.

Subbiah, M. (1997). Social construction theory and technical communication. In K. Staples & C. Ornatowski (Eds.), *Foundations for teaching technical communication: Theory, practice and program design* (pp. 53–65). ATTW Contemporary Studies in Technical Communication Series, Vol. 1. Greenwich, CT: Ablex.

Thralls, C., & Blyler, N. (2002). Cultural studies: An orientation for research in professional communication. In L. Gurak & M. Lay (Eds.), *Research in technical communication* (pp. 185–209). Toronto: Ablex.

Workman D. (2006, September 8). *Indian trade profits: World's third largest economy by 2040.* Suite 101. Retrieved June 29, 2008, from http://internationaltrade.suite101.com/article.cfm/indian_trade_profits

CHAPTER 4

Globalizing the Technical Communication Classroom: Killing Two Birds with One Stone

Emily A. Thrush and Angela Thevenot

As the world grows more globally connected, two challenges face the technical communication instructor. One involves introducing students native to the United States to the complexities of international and multicultural communication. The other concerns working with nonnative speakers of English and helping them develop their skills in English. We hope to address both in this chapter, with some suggestions for turning these challenges into an asset.

We are all aware of the numbers of nonnative speakers in our technical communication service courses. According to the National Science Board, in 2000 (latest figures available), 4,500 out of almost 60,000 graduates in engineering were foreign citizens, and 2,800 of 38,000 in computer sciences, for a total of 7% to 8% of undergraduate degrees in engineering and computer science. At the graduate level, international students received 45% of degrees in computer science and 38% in engineering. Not all international students are nonnative speakers of English, of course, but even those from India or other places where English is a primary language often have dialectal and cultural differences.

At the same time, there has been increasing interest in preparing native English-speaking technical writers for the global society of today and the future. The figures on the numbers of U.S. companies doing business globally and the numbers of Americans working for international companies have been cited elsewhere and do not need repeating. The premise of this collection of essays is that more attention needs to be paid to international and multicultural communication in the technical communication class, and more focus needs to be placed on the challenges facing international students writing for academic purposes in the U.S. academic culture.

The idea of dealing with both these challenges simultaneously led to the collaboration of the authors on a pilot project that will be described later. Emily teaches courses in the technical communication program: both the service course for engineering majors and the introductory course for technical writing majors, which is also required for engineering technology majors. She also teaches in the applied linguistics program and taught nonnative speakers of English for many years, in intensive English programs and in the service course in technical communication at Georgia Tech before coming to the University of Memphis. Angela is a graduate student in the doctoral program in applied linguistics and teaches in the Intensive English for Internationals program at the University of Memphis. What follows is our combined attempt to find effective ways of guiding international students to academic success and native students to professional success in the global community.

CHALLENGE 1: WORKING WITH NONNATIVE SPEAKERS OF ENGLISH

Nonnative speakers of English come to technical communication courses with a certain level of English language skills, depending on the entrance requirements for the particular university, but instructors often have unrealistic expectations of the fluency and accuracy those students can achieve in a relatively short period of time. For students in K–12 schools in the United States, the rule of thumb is that it takes 2 to 3 years for them to acquire basic interpersonal communication skills, but 5 to 7 years to gain the academic skills, especially in reading and writing, of their grade-equal native speaking peers, even when fully immersed in English for all school studies (Collier, 1987; Klesmer, 1994). International students in universities often had several years of English in school in their own countries, but that often means the equivalent of a few hours a week. They usually then attend an intensive program where they study English for 4 to 5 hours a day for 1 to 4 semesters before achieving the required score on the Test of English as a Foreign Language (TOEFL). These students usually gain much better skills in English than most Americans do in their foreign language studies, but such students are still not at native-like levels. Their English is often noticeably different from that of the native speakers in surface-level areas such as spelling, article and preposition usage, and verb forms; but also in ways that are more difficult for an instructor to identify. In a meta-analysis of research studies on the writing of international students enrolled in technical communication courses, Lehman, Nduna, van der Geest, and Winberg (2006) commented that researchers had

> an initial focus on issues of lexico-grammatical error, style and register inappropriateness and difficulties with the organization of text. The

> overwhelming majority of the recent research on ESL students' technical writing focuses on how the social practices of students' home-based primary and secondary discourse communities impact on the acquisition of technical writing competence in English. (p. 2)

The shift in research priorities indicates that researchers have identified the larger discourse issues as the most important; instructors, however, are often distracted from these by the perhaps more obvious issues of grammar and word choice. However, once identified, the issues of organization and discourse strategies are often more easily remedied because of the sheer number of possibilities for grammatical and spelling errors. With universities becoming more and more dependent on international studies, especially for programs in science and technology that attract fewer native speakers, we have to be realistic about what we can and cannot accomplish with these students.

WHAT WE CAN'T TEACH

Anyone can learn almost anything, given enough time, but in the restricted time of a semester, or an academic year, or even the 4 years of an undergraduate degree, there are limitations on what is possible. While the reasons are debated, it is generally accepted in the field of second language acquisition that nonnative speakers of a language rarely reach native-like fluency (Lenneberg, 1967, p. 158). Unless learners have significant exposure to native speakers before puberty, it is highly unlikely that they will ever sound like a native speaker, although they may achieve fluency and comprehensibility (Aitcheson, 1989; Coppetier, 1987; Scovel, 1988, and others). Other aspects of English that are highly problematic for nonnative speakers include the article system, which is especially difficult for speakers of languages that don't have articles, such as Chinese (Fen-Chuan, 2001), and idiomatic expressions such as phrasal verbs (Thrush, 2001). Studies on the effects of error correction on the acquisition of these grammatical features vary in results, but most indicate that drawing students' attention to the error results in some improvement in multiple draft situations but has little effect on the students' ability to self-correct (Ferris & Roberts, 2001; Truscott, 1996, 1999). While there are strategies that students can use to reduce the numbers of surface-level errors (peer review, use of checklists, spelling and grammar checkers, for example), it is unrealistic to expect nonnative speakers to produce flawless English.

Deep Culture

Anthropologists differentiate between surface culture (those things that are visible, readily explained, and fairly easily changed) and deep culture (those aspects of culture that embody deeply held values and beliefs and are therefore not easily changed) (Geertz, 2000; Williams, 1999). Surface culture in the technical communication classroom might include the formats to use for various types of documents and the procedures to follow in creating and submitting assignments. Elements of deep culture include the relationship between supervisor and employee (which affects tone, hedging and politeness strategies, and vocabulary choice), organizational values (which affect the choice of evidence in recommendations and proposals), and the way the society views written versus oral communication.

In academia, deep cultures affects such things as how a student enters and behaves in a classroom, how the student views collaboration versus independent work on assignments, and often, the students' writing style. In many cultures, sophistication of writing style is an important means of demonstrating one's class status and educational level; students may therefore resist instruction to write simple, clear sentences (Thevenot, 2007; Thrush, 1997).

CERTAIN GRAMMAR POINTS

There are some grammar points that are also very resistant to change through explicit teaching. For a second-language learner, there are two kinds of errors: interference from the native language and overgeneralization due to the inconsistencies of English (Scovel, 2001). An example of the second type of error is the student who has learned that "-ed" is the past tense marker in English and overgeneralizes to construct all past tenses with "-ed," resulting in sentences such as "I drinked and eated a lot last night." This, of course, is a fairly basic error, and most students have acquired the ability to use irregular verbs fairly well by the time they get to the technical communication course. Problems with prepositions persist much longer because, beyond the basic prepositions of place, they are often idiomatic. There's no general rule that tells the learner which preposition to use with particular adjectives and verbs, such as "appropriate" (appropriate to the national development policies? appropriate for the national development policies?) or frustrated (by? with? about?)—native speakers don't always agree.

Interference from the native language also affects the English of even very advanced learners. Interference can include transfer from the native language such as

- students who say "informations" because "information" is a countable noun in their languages and has a plural form;
- leaving a pronoun in an adjective clause such as "This algorithm was the one that we ran it" because that's the way adjective clauses work in their languages;

- using an infinitive rather than the simple form in a construction, such as "We can to see the results" because infinitives are one word in their languages and occur after verbs with the same meaning as "can."

Interference also happens when a structure doesn't exist in the native language, making it very difficult to acquire in a second language. English speakers learning Turkish often find it very hard to remember when to use the inclusive "we" (when the speaker is referring to himself/herself and the listener) and when to use the exclusive "we" (when the speaker is referring to himself/herself and some other people not including the listener). Similarly, speakers of languages without articles ("the" and "a" in English) generally find it almost impossible to master the article system in English. Even speakers of languages with articles make mistakes because article usages is often idiomatic and arbitrary (consider the standard use of "the" with road numbers on the west coast, as in "Take the 5 to get to San Diego" versus the zero article in most of the rest of the country as in "Take 240 to the Poplar exit.").

Although certain culture or grammar points cannot be taught and must be acquired experientially by English for Academic Purposes (EAP) students, other aspects of the language and culture can and should be taught in order to encourage academic performance and academic acculturation.

WHAT WE CAN TEACH

Grammar

There are two issues here: what grammar to teach and how to teach it. We have already discussed the fact that it may be a waste of time to focus on aspects of the language that are not likely to be acquired by a nonnative speaker and do not interfere with communication. It is also well-established that explanation of grammar rules does not necessarily result in acquisition of the correct forms (this is true for native speakers as well as Non-Native Speakers (NNS), as the debates on the role of grammar correction in the composition class have demonstrated).

To improve academic language proficiency,

> Students need a two-pronged approach to grammar and language instruction. They need to learn to see and edit the errors they produce. At the same time, they need guidance in expanding their grammatical and lexical repertoire based on how language works in the content texts. Let students' written production drive your decisions about what to teach. (Brinton & Holten, 2001, p. 250)

Learning in the context of their own written production is always more effective than decontextualized exercises involve rewriting individual sentences in a workbook. Nor does marking all the errors in an assignment with cryptic codes ("sp," "s/v agree") result in much learning. What does seem to help many students is an error-analysis approach in which a particular type of error is underlined or circled, and the student has to determine what the error is and how to correct it.

Another approach is something called "input processing" (VanPatten, 1996) in which the students' attention is drawn to correct forms of grammar by pointing them out in context, and having the students process the grammar in order to make some kind of decision. For example, students might be given sentences such as "Life can be hard" and "The life of an engineer can be hard" and asked to decide whether the reference is to a specific person or group or a general statement applicable to all people. This approach can help students acquire internalized rules for the use of "the" with abstract nouns.

It is true, however, that students often want explicit, systematic instruction in grammar (Brinton & Holten, 2001). Perhaps that is because of their prior experiences with language instruction. Students often ask for grammar correction, although research has shown that too much emphasis on accuracy impedes their development of fluency, in both writing and speaking. Failure to meet their expectations for correction often results in feelings of frustration and a sense that the teacher is not doing his or her job. It can be difficult to balance these expectations with a more developmental approach to grammar acquisition and with the time constraints of the classroom and the academic term, but helping students to see that some aspects of language are more critical for communication helps, as does a focus on grammar in context. Brinton and Holten (2001) give a number of effective ways to focus on grammar and vocabulary within a specific content area.

One problem is that many instructors of technical communication classrooms, especially teaching assistants, are uncomfortable with grammar instruction featuring teacher-fronted explanations and form-focused exercises, partly because they themselves are not aware of the trouble-spots in grammar or how those structures function in English (most, for example, cannot explain a modal, describe the rules for article usage, or give the appropriate uses of the present perfect aspect) (Brinton & Holten, 2001, p. 247). They may also feel that NNS students should be using perfect, American English before entering their classrooms, despite the fact that second language research indicates that it takes 7 to 9 *years* of intensive immersion experience for most NNS to acquire academic language skills. This implies that teacher-preparation programs should promote teachers' essential language awareness so that they can better apply theory to practice (Ramanathan, 2002).

Collocation

Collocation activities teach both grammar and vocabulary through the use of authentic materials that illustrate how words and structures occur in a particular discourse community. Current authentic materials can be particularly engaging to native speakers and NNS students alike, as they are more contemporary and relevant than most textbook materials. By introducing materials drawn from the media, which are about current events in the real world and happening in familiar places—characterized by the genuineness of time, location, and people—students may find it easier to relate the events to their own experience or knowledge and be able to appreciate the use of language in these materials (Kwok & Choi, 1995).

Relevant and engaging materials encourage NNS students to enter the authentic English conversation of their professional fields. Used as input pushing tools (VanPatten, 1996, p. 93) and agents of motivation, these materials serve to enhance the opportunity for international students to synthesize information and apply it via critical-thinking skills to everyday conversation. It is this conversation that overlaps cultures, resulting in an international learning environment that is rich in intercultural dialogue at spoken and discursive levels.

Academic Culture

Writing students learn to write *for* a specific audience. For native speakers of English studying at an American university, this perspective involves applying rhetorical practices learned as academic writers to a designated genre. These students are writing in a culture that shares thought patterns, discourse expectations, as well as academic and sociocultural beliefs. International writing students, on the other hand, have a more challenging position. They learn to write *for* a specific audience and, at the same time, are learning to write *in* the academic cultural environment in which the audience exists.

According to Flowerdew and Miller (1995), "Academic culture refers to values, roles, assumptions, attitudes, patterns of behaviour, etc., which may differ across cultures." These implicit differences present challenges for international students; challenges which require time and experience to negotiate. On a more explicit level, academic culture involves coachable issues: plagiarism, genre-specific organization (discourse structure, formatting and the importance of "curb appeal"), examinations procedures and evaluation, grammar pedagogy, and the like. In these areas, learners can be coached to produce work that will satisfy expectations that are a part of the academic surface culture. At the same time, it is important to respect, value, learn from, and build on the identity of each student.

PLAGIARISM

One cultural matter not addressed in the profile on cultural attitudes and values (see Appendix A) used in our study (described later) is the issue of plagiarism, or intellectual ownership. R. R. Jordan (2005) addresses complications surrounding the idea of intellectual ownership in *English for Academic Purposes: A Guide and Resource Book for Teachers*:

> In a number of countries, particularly in Asia, there is considerable respect for the printed word and those in authority. Consequently, it is quite normal for students to quote from authorities/books without feeling the need to acknowledge the source; nor is it necessarily expected. Criticism of published works is unacceptable as it displays disrespect for authority. Consequently, students will often present extracts from books or articles in an uncritical way. (Cortazzi, 1990)

This view is in sharp contrast to the American position that the printed word is owned by the writer and consequently, protected—to copy is theft. Although in business and industry, text belongs to the organization, and technical writers on the job frequently start with text written by another, often anonymous writers shape it, adapt it, and build on it without any kind of acknowledgment of the original source. In the classroom, we typically apply the more standard academic attitudes about using the words of other. Perhaps in the future, American academia will adopt the Bakhtinian (1986) notion that all discourse is generated from prior discourse and align itself with the global consensus. For the moment, however, it is important that the American university definition of and attitude toward plagiarism be explained and discussed with the nonnative writer. The policy at our university reads as follows:

> Plagiarism is a serious academic offense. The University of Memphis regards plagiarism as academic dishonesty. Consequences of plagiarism include failing an assignment, receiving a lower course grade, and even failing a course. According to the University of Memphis Code of Student Conduct "[t]he term 'plagiarism' includes, but is not limited to, the use, by paraphrase or direct quotation, of the published or unpublished work of another person without full or clear acknowledgment. It also includes the unacknowledged use of papers or other academic materials." (http://www.people.memphis.edu/~jaffairs/csc/definitions.htm: Definition 15)

To the native speaker (NS), this message clearly explains the consequences of using material unlawfully as one's own. It is often less clear, even to the native speaker, how much of the original language can be retained in a paraphrase, and when the line has been crossed into plagiarism. For example, if the student is writing about atomic particle acceleration, those three words would certainly be expected to be used together. What about phrases such as "atomic particle acceleration paths," "artificial atomic particle acceleration," or "sub-atomic particle acceleration?" (all found by a Google search). At what number or combination of words has the student borrowed too much from the original? The definition of plagiarism is often less clear than we pretend, even to the native speaker. To the NS, even the policy may be unclear. What does the phrase "but is not limited to" mean? What is being left out—not said? To the international student who does not share the concept of intellectual ownership, this policy is not only confusing, but morally perplexing. A brief discussion about the American academic culture being rooted in puritan ethic may help to shed light on the resulting moral responsibility carried by the student. Failure to address this topic in a globalized classroom could lead to significant misunderstanding and possible academic sanctions for the international student. One participant in our project experienced such a misunderstanding at our university.

The Asian student was enrolled in a concurrent World History class, for which she was instructed to write an expository essay. All written work for this class was to be submitted to Turn-it-in.com,

> an internet based program created by iParadigms, LLC, which is most renowned for its plagiarism detection service. Institutions (typically universities and high schools) can buy licenses to use the plagiarism detection service. Teachers may then submit essays to the Turnitin website to check for plagiarism, or—as is most commonly practiced—students submit essays directly to the website themselves. (http://en.wikipedia.org/wiki/Turnitin)

The student, from South Korea, was instructed to write on Confucian cosmology. Coming from an Asian culture that shares much of the same ideas as those discussed in her paper, she assumed that the information that she researched and submitted would fall under the realm of "common knowledge," just as an American student would be able to write about George Washington being the first president of the United States without citing a source. When the paper was selected by the monitoring software as being plagiarized, the student was baffled and extremely upset.

Many times, particularly in the fields of science, incidents such as this arise because students who have taken limited basic composition coursework and are lacking in writing-process skills are given assignments consisting of substantial writing over a short period of time. Pressure to perform and commitment to complete coursework in a timely manner overshadow the need to follow misunderstood moral convictions.

Using this situation as a teaching point, one of the authors of this chapter (Angela) prepared a relevant real-time lesson plan on an alleged plagiarism incident in the business world. At the time of the course, major newspaper publications were reporting on Raytheon executive CEO William Swanson's denial that he plagiarized a book on rules of management written by W. J. King 60 years earlier (Raytheon chief says he didn't plagiarize, 2006).

Following a warm-up discussion on the idea of plagiarism and its applicability to areas outside the academic environment, students were given an article to read from *USA Today*, titled "Raytheon chief says he didn't plagiarize." They were next given a copy of *The Unwritten Laws of Engineering* (King, 1944), printed in parallel with *Swanson's Unwritten Rules of Management* (see Appendix B).

Students were then put in pairs and instructed to use contrastive analysis to deductively determine grammatical and linguistic differences and similarities in the two texts. They then were asked to discuss their findings and make a decision as to the justification of plagiarism charges. The decision, based on empirical evidence (including replacement of dated language, change in determiners, and many cases of word-for-word copying), was unanimous: clearly an act of plagiarism had been committed. The day after this activity, one student, on his own, found a statement issued by the CEO on the Raytheon Web site demonstrating that students are engaged by authentic, relevant, real-time tasks. The activity offered the students an opportunity to experience authentic technical writing samples, and at the same time, to become part of current business conversation.

CHARACTERISTICS OF ACADEMIC WRITING

Postcolonial international students are aware of differences in discursive patterns and rhetorical strategy. "International students come into the university from a range of educational points of entry. This contributes to their hybridity and multiplicity" (Kumar, 2004, p. 145). In her dissertation, *The Discursive Representation of International Undergraduate Students: A Case Study of a Higher Education Institutional Site*, M. K. W. S. Kumar (2004) describes the educational histories of some of the students:

> Mothudi says that Botswana has a British system of education that continued after independence. Nimali has also studied under a British system in Sri Lanka. In Teresa's situation, Hong Kong was under a British government before 1997 so they had a British system of education, but the Cantonese language was used to explain it. Sarojini has been to Christian, Islamic and Hindu schools. For other students their learning has been in the language spoken by the community. Sherly, for instance, says that her primary education was conducted in Bahasa Malayu. Fen Yi's was conducted in Vietnamese and Rebecca Hasnun's was conducted in Indonesian. (pp. 145–146)

This awareness often causes confusion when these students face writing assignments in an Anglo-Saxon university environment. As second language students with knowledge of a range of discourses, they struggle to write with coherence under academic discursive constraints of the host institution. For some students (e.g., native speakers of Asian languages or of Arabic) who do not share the target language writing script, mechanical difficulty impairs writing speed. These problems can be addressed with form-focused direct instruction by the writing instructor. Behaviorist in nature, form-focused instruction is product oriented and at the same time reader based. As such, it enables the students to generate work meeting expectations and values of the academic culture in which it is written. The teaching challenge is determining exactly what expectations of the academic culture we can teach and how we can address the writing needs of nonnative speakers in a way that is appropriate to them.

Curb Appeal: Fact: a student's work is initially judged by its appearance. An instructor will make an initial, albeit subliminal, decision on a particular paper strictly based on its apparent appropriateness to task. This approach is even more true in a workplace situation. Incorrect formatting often detracts from the merit deserved of a work's content. While a student must produce knowledge-based content on his or her own, an instructor can direct the overall appearance of a writing project, teaching the international student to properly format specific genres to meet the expectation of their academic fields and of the professional community they will be joining. A way to do this is by using textual modeling to demonstrate the discourse structures and organizational strategies found in the wide variety of academic and professional genres; genres that are culturally, socially, and disciplinarily determined by the dominant norm. They are composed of specific content and are written for a target audience.

Genre Organization: The short answer question, the essay question, the argument, the documented research essay, the scientific research project, the thesis, and the dissertation are all genres used in the American academic community. E-mail, the memo, the business letter, the professional report, and the journal article are all genres used in the American business environment. Each consists of a predictable, contextualized framework that can be "read" by the student and used as a template for producing like work. Using authentic samples and encouraging discourse analysis of the framework—the setting, the purpose, the content, the intended audience, the relationship between the writer and the readers, the discourse community's expectations, the structure and language, and so on (Grabe & Kaplan, 1996)—of any given genre, the writing instructor can inform the student of expectations of that genre, remove stress, and help the writer achieve coherence in his or her writing. For example, students in technical and scientific fields will need to produce research reports. Teaching a generic IMRAD (Introduction, Methodology, Research, and Discussion) form will give them a working knowledge of a genre that will fulfill writing requirements in their field. This structure is found in every type of scientific research article and is by far the most common used in writing the dissertation, with almost 80% of dissertations using a variant of the IMRAD Format (Paltridge, 2002). With genre awareness, the student is empowered to autonomously perform assigned tasks.

A Radical Proposal

The assumption has always been that international students need to be able to produce flawless American English; that if we accepted anything less, we would be lowering our standards. If we didn't count off for every article and comma mistake, regardless of the mastery of the subject material, we would be violating the interests of the academy and the student. We would like to suggest a different approach.

English no longer belongs to any of the countries in which it is the primary language. It is difficult to estimate the numbers of people using English and their proficiency with it. Various estimates have put the numbers of native speakers at about 350 million and nonnative speakers who use English on a fairly regular basis at about 1,800 million (Crystal, 1997, p. 61) for a total of around 2 billion. The closest other language in number of speakers is Mandarin Chinese (Putonghua) with about 1 billion, but most of those reside in one country. English, on the other hand, is used to some extent in all countries on all continents.

While the degree to which English is used is impossible to estimate (some nonnative speakers conduct most of their professional life in English, while others may use it for the occasional business transaction, for travel, or for international conferences, etc.), the important factor is that there are many more nonnative speakers using English than native speakers, and they are mostly using it with each other rather than with native speakers of British or American or Australian or any other specific dialect of English. This situation is easily seen in any international venue: travelers from all over the world are more likely to be using English with hotel desk clerks or restaurant staff than the native language of whatever country they are in. A recent visit to the Vatican revealed that a portion of the Mass said over loud speakers to the thousands of visitors was spoken only in English, with the assumption that more of the throng would understand that language than any other, including Italian, Latin, or the native language of the priest saying the Mass.

In response to this phenomenon, what has emerged is something linguists call "World English": a variety of the language lacking most of the features specific to any one dialect and without some of the features common to those dialects that are the most problematic for second-language learners to master. This includes pronunciation features such as the "th" sounds (which few other languages have), the articles "a" and "the," and many of the idiomatic expressions, especially those containing prepositions, such as the phrasal verbs (Kachru, 1985; Thrush, 2001). So, for example, second-language speakers are more likely to say, "It is equipped with another rocket motor that is used to *accelerate* space craft" than "It is equipped with another rocket motor that is used to *speed up* the space craft"[1] (see Table 1).

An examination of many technical journals will yield examples of texts written by nonnative speakers of English in which the article and preposition use is nonstandard for any dialect, but these "errors" not only do not interfere with communication, but may actually make the text more readable for other nonnative speakers.

[1] This example found at http://teams.kipr.org/2006/06-0003/research/equipment.htm, May 13, 2006.

Table 1.

Characteristic of World English	Example
Use of sense verbs in the progressive	"I am hearing a noise": "I am seeing GRAY BAR again in the Google Toolbar."[a]
Lack of inversion in questions	"What they would like to do in the future?" "What food you can't eat?"[b]
Lack of "do" insertion in questions	"Where you go yesterday?"[c]
Use of "the" with abstract nouns	"It criticises people's relationship with *the nature* and the outlook patriarchal society has on women."[d] "Do you believe that the chocolate is better than a kiss?"[e]

[a]Posted May 15, 2006, by a Turkish speaker on SEO Chat Forums/Google.
[b]Found November 2, 2007, at http://forums.eslcafe.com/student/viewtopic.php?t=23249
[c]Found November 2, 2007, at http://emptybottle.org/glass/2005/11/language_drainage.php
[d]This example found May 14, 2006, at http://www.kaltio.fi/index.php?499
[e]Found November 2, 2007, at http://forums.eslcafe.com/student/viewtopic.php?t=24314

Our radical proposal, then, is that it may be more important for native speakers of English to learn to communicate in a manner that is understandable to nonnative speakers of English than the other way around. That was the basic premise of Thrush's article on Plain English (2001): the "short, everyday" words advocated by Plain English groups as being easier for native speakers to process and comprehend in fact made comprehension much more difficult for nonnative speakers. The milder version of this argument is that, at the very least, it's not worth worrying about article usage in texts written by nonnative speakers because they're not going to master idiomatic usage of articles anyway, certainly not in the semester they are in your class; and they don't need precise use of articles to communicate, especially with the other almost 2 billion nonnative speakers. Tony Silva, editor of the *Journal of Second Language Writing,* calls this "writing with an accent" (2006). That is, we would not expect a world renowned scientist from China, when giving a speech at an international conference, to speak English without an accent. We expect pronunciation differences and are willing to compensate, to a certain extent, for sound differences, missing sounds, and stress and intonation differences. We are, in fact, usually impressed when a nonnative speaker is articulate and comprehensible in English. Somehow, however, we expect "accentless" production in written English.

On a broader level, we have to consider also whether we should be insisting on the rhetorical patterns typically taught in the technical communication classroom. Land and Whitley (2006) make the point in regard to the composition classroom that we may not be doing ourselves or our discipline any favor by limiting the students to a set of cohesion and organization strategies predetermined as acceptable. They quote McCannell and McCannell (1982) who said, "culture that reproduces itself as a series of endless mirrorings, yet adds nothing to either the original 'natural culture' or the original 'image' of it, is literally the death of the culture" (p. 325). Land and Whitley also cite research showing that while native speakers rate essays written by other native speakers much higher than those written by nonnative speakers, the nonnative speakers rated both groups of essays about equally. That is, nonnative speakers found papers using nontraditional rhetorical patterns equally comprehensible and equally well-organized when compared with those deemed acceptable by native speakers (pp. 326–327). Their conclusion is that nonnative speakers are able to accommodate a much wider range of rhetorical patterns and strategies; that it is the native speakers who are limited in their ability to grasp meaning from texts that they might encounter in the course of international business or research. Also, as Canagarahah (2001) points out, "the discourses of post-colonial subjects are multiple, hybrid and overlapping. . . . The challenging question we need to explore is: how do students negotiate these conflicting and overlapping discourses to create coherence in their academic texts?" (p. 121).

That brings us to the issue of how to prepare both native and nonnative speakers for the global use of English.

CHALLENGE 2: INTRODUCING NATIVE SPEAKERS TO INTERNATIONAL COMMUNICATION

The literature already contains a number of ways of raising awareness of writing for culturally and linguistically diverse audiences. Deborah Andrew's textbook, *Technical Communication in the Global Community* (2000), offers case studies and documents for students to analyze. In Maylath and Thrush (1999) and Thrush (1997), the authors suggest several ways of incorporating a focus on international communication into a general tech comm class, as well as some of the linguistic and cultural elements to include. At Georgia Tech, TyAnna Herrington conducts the Global Classroom Project, in which students at Georgia Tech and at a European university in St. Petersburg, Russia, participate together in digital classes and collaborative projects (http://www.lcc.gatech.edu/~herrington/gcpdescription/gcpdesc.html).

Creating Awareness of Cultural Genre

We would like to propose another approach that we piloted in the spring of 2006. This approach attempts to both internationalize the technical communication class and to build the skills of international students by bringing the two groups of students together for specific targeted activities. As Crane, Galvanek, Liamkina, and Ryshina-Pankova (2002) commented in a study of perceptions of genre, "Most—if not all—of our informants see genre as a culturally, socially and historically determined and situated practice. To quote one of the informants: 'the awareness of genre is a part of cultural literacy.'" Building cultural literacy, then, includes encouraging students to recognize multiple literacies.

The Project

To do this, we decided to capitalize on a resource readily available at most large U.S. universities: international students learning English. Emily was teaching the introductory course in technical communication, required for professional writing major and engineering technology students. Angela was teaching the academic-preparation course in the Intensive English for Internationals program. Most universities have an intensive English program such as this one, often located in the Extended Program or International Programs division. Students in these programs are primarily preparing to attend a U.S. institution of higher education, and the majority are planning to major in math, science, computer science, or engineering. (At other schools, where technical writing majors and engineering students are in the same class, it may be possible to do the activities below without going to the individualized educational program, but we did not have any nonnative speakers in the intro technical communication course.)

Our pilot of this activity ran into some snags, primarily because the technical communication class was entirely online, and this project had not been sufficiently incorporated into the course syllabus. However, the following describes the steps as we planned them.

The Process

1. Both sets of students are prepared by their respective teachers. The native speakers read an article by Thrush (1993) on international communication. The international students take part in a discussion of globalization and cultural differences. Both sets of students read a handout on high- and low-context cultures to set a base level of understanding of cultural differences. All students in both groups individually complete a profile on cultural attitudes and values (see Appendix A).
2. Next, the two groups meet and are paired up. The pairs review the cultural profile, discussing the places where their answers differed. A discussion with the whole group ensues, in which the teachers elicit more information from both groups about how the culture values highlighted in the profile operate in their cultures.

3. Students in both groups complete a journal entry reflecting on what they have learned. The journal entry includes a t-bar graph with their comparisons of the cultures represented in the discussion group.
4. Students meet outside class with their partners later in the semester to do the brainstorming and planning for a recommendations report. The topics for the report are campus issues that both sets of students have experienced. Together, they work through an analysis of audience and purpose, develop a plan for organizing the report, and discuss what evidence to include. The purpose of this is to give both sets of students the chance to see that they have different concepts of what their audience might be interested in, what pattern of organization is appropriate, and what types of evidence are valued by their readers.
5. Technical communication students write their recommendations reports separately. They are under no obligation to use any of the suggestions from their international partners.
6. Both sets again complete a journal entry reflecting on what they have learned.

The Results

In our pilot of this activity, a few interesting things emerged. First, the international students all pointed to a generation gap in their countries. On the profile survey, the "a" answers are considered to be a reflection of mainstream U.S. culture. The "b" answers represent various other cultures. These international students, who ranged in age from 19 to 24, all indicated that their parents would probably have chosen more "b" answers than they did. To some extent, this difference might reflect a change in their cultures, with more emphasis on individual achievement than in the past, but to some extent it also reflects both their age and the fact that the socioeconomic and educational backgrounds of students who come to the United States to study are not necessarily representative of the majority of young people in their country. However, these students are probably more representative of those who will be in management positions in their countries or involved in international research in the sciences.

The native speakers all expressed surprise that the cultural differences were not as great as they expected, but realized that this had something to do with shared educational experiences and aspirations. One commented, "One question that made this obvious to me was whether helping a fellow student on a test was cheating. We all agreed that it was because of our similar experience with university policy."

In terms of analyzing audiences, the native speakers gained some understanding of different ways in which relationships between levels within an organization differ. One said, "Some of the answers reflected the work ethic of the different cultures. For example, in Africa they believe an employee should wait for directions from a supervisor and in Korea and the United States, we find being self-motivated and taking the initiative to get work done a good quality in an employee. Another difference in culture was how you could obtain success in the work place. In Africa and Korea, family ties and friendship have a great influence on your ability to get certain jobs, whereas in the United States, we tend to hold the belief that if we work hard and gain experience, we can qualify for any job."

CONCLUSION: KILLING TWO BIRDS

Neither of the challenges we have discussed is going to yield easily to any particular instructional strategy; there is no "magic bullet." What we have tried to suggest here is a twofold approach, which includes changing our expectations that international students can, or should, be expected to conform to the model of the native speaker student, and that native and nonnative speakers can become resources for each other in learning about language, culture, and the skills needed for their professional lives. We also see a need for greater awareness of linguistic concepts and cultural differences for all technical communication instructors, and a greater inclusion of these areas in teacher-preparation programs. Academic discourse is not a "homogenous entity" (Adler-Kassner & Estrem, 2005). "Students have to engage with a range of discourses in academic writing with which they need to engage critically in order to produce a variety of genres" (Sheridan, 2005). There are proposals currently under consideration in the U.S. Congress to create new visa categories for the specific purpose of attracting more international students in the areas of science and technology because of the needs of both universities and corporations, so the issue of international students and communication with international audiences is only going to become more critical in technical communications programs. We hope that these will truly be seen as opportunities in the future, rather than challenges or problems.

Appendix A follows

APPENDIX A
Profile on Culturally Determined Attitudes and Values

Directions: For each item, choose either "a" or "b," whichever comes closest to describing your own personal values. Record "a" or "b" in the box to the left.

	1a. Hard work will accomplish most objectives. 1b. Wisdom, luck, and fate are basic to success.
	2a. Communication should be frank and direct. 2b. Communication should be polite and indirect.
	3a. Commitments should be honored. 3b. Commitments signify intention and may be superseded by a conflicting request or need.
	4a. Time should be effectively and efficiently used. 4b. Schedules should be viewed in relationship to other priorities.
	5a. An individual's work tells a lot about him/her. 5b. Individual identity isn't defined by work or accomplishments.
	6a. Success is earned by ability, experience, and hard work. 6b. Family ties and friendships determine mobility.
	7a. Teachers should consult with parents, who can contribute useful information. 7b. Decisions should be made by those in authority and others need not be consulted.
	8a. Competition stimulates high performance. 8b. Competition leads to disharmony.
	9a. One should do whatever is necessary to get the job done. 9b. Some work may be below one's dignity.
	10a. Change is healthy. 10b. Traditions should be preserved by observing rituals and customs from the past.
	11a. Both persons and systems should be evaluated. 11b. Personal evaluations should never be negative because they might cause a "loss of face."
	12a. Helping a fellow student on a test is cheating. 12b. One should help friends with school work, even tests.
	13a. Present success and satisfaction are important. 13b. Materialism isn't important; spirituality is.
	14a. Employees should be self-directed and take initiative in the work. 14b. Employees should wait for directions from the supervisor.
	15a. An individual should do everything possible to achieve his/her goals and aspirations. 15b. The group is more important than the individual.

Total number of "a"s ______ Total number of "b"s ______

Source: Center for Applied Linguistics

APPENDIX B:
Comparison of King and Swanson Texts

Sixty years before Raytheon CEO William Swanson published his *Swanson's Unwritten Rules of Management*, W. J. King (1944), a UCLA engineering professor, published *The Unwritten Laws of Engineering*. King's book contained about half of the rules that appeared in Swanson's book.

King, 1944: "Cultivate the habit of 'boiling matters down' to their simplest terms."
Swanson, 2004: "Cultivate the habit of boiling matters down to the simplest terms: the proverbial 'elevator speech' is the best way."

King: "Do not get excited in engineering emergencies—keep your feet on the ground."
Swanson: "Don't get excited in engineering emergencies: keep your feet on the ground."

King: "Cultivate the habit of making brisk, clean-cut decisions."
Swanson: "Cultivate the habit of making quick, clean-cut decisions."

King: "Promises, schedules, and estimates are necessary and important instruments in a well-ordered business."
Swanson: "Promises, schedules and estimates are important instruments in a well-run business. You must make promises—don't lean on the often-used phrase: 'I can't estimate it because it depends on many uncertain factors.'"

King: "In dealing with customers and outsiders remember that you represent the company, ostensibly with full responsibility and authority."
Swanson: "When interacting with people outside the company, remember that you are always representing the company. Be especially careful of your commitments."

King: "Be extremely careful of the accuracy of your statements."
Swanson: "Be extremely careful in the accuracy of your statements."

King: "Don't overlook the fact that you're working for your boss. Whatever the boss wants done takes top priority."
Swanson: "Don't overlook the fact that you are working for a boss. Keep him or her informed. Whatever the boss wants, within the bounds of integrity, takes top priority."

King: "Be as particular as you can in the selection of your boss."
Swanson: "Work for a boss to whom you can tell it like it is. Remember, you can't pick your family, but you can pick your boss."

King: "Strive for conciseness and clarity in oral and written reports."
Swanson: "Strive for brevity and clarity in oral and written reports."

King: "Don't be timid—speak up—express yourself and promote your ideas."
Swanson: "Don't be timid: speak up, express yourself and promote your ideas."

King: "Confirm your instructions and the other fellow's commitments in writing. Do not assume that the job will be done . . . "
Swanson: "Confirm the instructions you give others, and their commitments, in writing. Don't assume it will get done!"

King: "However menial and trivial your early assignments may appear give them your best efforts."
Swanson: "However menial and trivial your early assignments may appear, give them your best efforts."

REFERENCES

Adler-Kassner, L., & Estrem, H. (2005). Reaching out from the writing classroom. In T. Kostouli (Ed.), *Writing in contexts: Textual practices and learning processes in sociocultural settings* (pp. 229–246). Amsterdam: Springer Netherlands.

Aitchison, J. (1989). *The articulate mammal: An introduction to psycholinguistics.* London: Routledge.

Andrews, D. C. (2000). *Technical communication in the global community* (2nd ed.). Upper Saddle River, NJ: Prentice-Hall.

Bakhtin, M. M. (1986). *Speech genres and other late essays.* Austin: University of Texas Press.

Brinton, D., & Holten, C. A. (2001). Does the emperor have no clothes? A re-examination of grammar in content-based instruction. In J. Flowerdew & M. Peacock (Eds.), *Research perspectives on English for academic purposes* (pp. 239–251). Cambridge, UK: Cambridge University Press.

Canagarajah, S. (2001). Addressing issues of power and difference in ESL academic writing. In J. Flowerdew & M. Peacock (Eds.), *Research perspectives on English for academic purposes* (pp. 117–131). Cambridge, UK: Cambridge University Press.

Collier, V. P. (1987). Age and rate of acquisition of second language for academic purposes. *TESOL Quarterly, 21,* 617–641.

Coppieters, R. (1987). Competence differences between native and near-native speakers. *Language, 63,* 544–573.

Cortazzi, M. (1990). Cultural and educational expectations in the language classroom. In B. Harrison (Ed.), *Culture and the language classroom* (pp. 54–65). Hong Kong: Modern English Publications and the British Council.

Crane, C., Galvanek, J., Liamkina, O., & Ryshina-Pankova, M. (2002, March 9). *Genre: Where art thou?: Tracing the role of genre in the foreign language curriculum.* Paper presented at the UC Consortium Conference on Language Learning and Teaching, Irvine, CA: (ERIC Document Reproduction Service No. ED475015).

Duenas, M. (2004). A description of prototype models for content-based language instruction in higher education. Retrieved April 2006, from http://www.publicacions.ub.es/revistes/bells12/PDF/art04.pdf

Fen-Chuan, C. L. (2001). The acquisition of articles by Chinese learners. *Second Language Studies, 20*(1). Retrieved April 2007, from http://www.hawaii.edu/sls/uhwpesl/on-line.cat.html

Ferris, D., & Roberts, B. (2001). Error feedback in L2 writing classes: How explicit does it need to be? *Journal of Second Language Writing, 10,* 161–184.

Flowerdew, J., & Miller, L. (1995). On the notion of culture in L2 lectures. *TESOL Quarterly, 29*(2), 345–374.

Flowerdew, J., & Peacock, M. (Eds.). (2001). *Research perspectives on English for academic purposes*. Cambridge, UK: Cambridge University Press.

Geertz, C. (2000). *Interpretation of culture.* New York: Basic Books.

Grabe, W., & Kaplan, R. (1996). *Theory and practice of writing: An applied linguistic perspective*. London: Longman.

Jones, D. (2006, April 25). Raytheon chief says he didn't plagiarize. *USA Today.* Retrieved from http://www.usatoday.com/money/companies/management/2006-04-24_raytheon-ceo-responds

Jordan, R. R. (2005). *English for academic purposes.* Cambridge, UK: Cambridge University Press.

Kachru, B. B. (1985). Standards, codification and sociolinguistic realism: The English language in the outer circle. In R. Quirk & H. G. Widdowson (Eds.), *English in the world: Teaching and learning the language and literatures* (pp. 11–30). Cambridge, UK: Cambridge University Press.

Klesmer, H. (1994). Assessment and teacher perceptions of ESL student achievement. *English Quarterly, 26*(3), 5–7.

Kumar, M. K. W. S. (2004). *The discursive representation of international undergraduate students: A case study of a higher education institutional site.* Unpublished dissertation, School of Education, University of South Australia, Adelaide.

Land, R. E. Jr., & Whitley, C. (2006). Evaluating second-language essays in regular composition classes: Toward a pluralistic U.S. rhetoric. In P. K. Matsuda, M. Cox, J. Jordan, & C. Ortmeier-Hooper (Eds.), *Second-language writing in the composition classroom: A critical sourcebook* (pp. 324–332). Boston: Bedford/St. Martin's.

Lehman, B., Nduna, J., van der Geest, T., & Winberg, C. (2006). A meta-analysis of the teaching of technical writing students for whom English is not a first language. Retrieved April 2006, from http://www.ncte.org/groups/cccc/highlights/120214.htm

Lenneberg, E. H. (1967). *Biological foundations of language*. New York: Wiley.

Maylath, B., & Thrush, E. A. (1999). Café, thé ou lait? How shall we train technical communicators? In P. Hager & H. Scheiber (Eds.), *Managing global discourse* (pp. 233–254). New York: John Wiley & Sons.

National Science Board. (2000). Science and engineering indicators 2004. Retrieved from http://www.nsf.gov/statistics/seind04/c2/c2s3.htm

Paltridge, B. (2002). Thesis and dissertation writing: An examination of published advice and actual practice. *English for Specific Purposes, 21*(2), 125–143.

Ramanathan, V. (2002). *The politics of TESOL education: Writing, knowledge, critical pedagogy*. New York: RoutledgeFalmer/Taylor and Francis.

Scovel, T. (1988). *A time to speak: A psycholinguistic inquiry into the critical period for human speech*. New York: Newbury House.

Scovel, T. (2001). *Learning new languages: A guide to second language acquisition.* Boston: Heinle & Heinle.

Sheridan, V. (2005). [Review of the book *Writing in Contexts: Textual practices and learning processes in sociocultural settings,* (T. Kostouli, ed.)]. Amsterdam: Springer Netherlands. In Linguist List, April 17, 2006. Retrieved April 2007, from http://linguistlist.org/issues/17/17-1237.html#1

Silva, T. (2006). *Writing across borders.* [Video.] Oregon State University. Available at http://cwl.oregonstate.edu/wab/

Thevenot, A. (2007, March 23). Uncovering academic culture in EAP content classes. *Teachers of English to speakers of other languages* (TESOL). Seattle, Washington.

Thrush, E. A. (1997). Multicultural issues in technical communication. In K. Staples & C. Ornatowski (Eds.), *Foundations for teaching technical communication: Theory, practice, and program design* (pp. 161–178) Greenwich, CT: Ablex.

Thrush, E. A. (2001). Plain English vocabulary and international audiences. *Technical Communication, 48*(3), 290–296.

Truscott, J. (1996). The case against grammar correction in L2 writing classes. *Language Learning, 46,* 327–369.

Truscott, J. (1999). The case for "the case against grammar correction in L2 writing classes": A response to Ferris. *Journal of Second Language Writing, 8,*111–122.

Turn-it-in.com. (n.d.). Retrieved May 31, 2006, from http://en.wikipedia.org/wiki/Turnitin

VanPatten, B. (1996). *Input processing and grammar instruction in second language acquisition.* Greenwich, CT: Ablex.

Williams, T. R. (1999). *Cultural anthropology.* Upper Saddle River, NJ: Prentice-Hall.

Wong, V., Kwok, P., & Choi, N. (1995). The use of authentic materials at tertiary level. *ELT Journal, 49*(4), 318-322.

SECTION II

Curricular Perspectives: Designing and Developing Courses and Programs in Intercultural Communication

CHAPTER 5

In memory of our friend and colleague Victoria Mikelonis

Incorporating "Shock and Aha!" into Curriculum Design: Internationalizing Technical Communication Courses

Shelley L. Smith
and Victoria M. Mikelonis

As Friedmann (2005) so aptly put it, the world is rapidly becoming flat. By making the access to, and the exchange and communication of, information a global concern, technology has become the driving engine behind this rapid change from cultural and economic isolation to integration. The Internet has made it as easy and as cheap for someone in New Delhi to access information at Harvard as it is for a student in Boston. As a result, interculturally competent technical communication professionals will be increasingly in demand as people struggle to cope with the communication concerns that result from this global intersection of cultures.

How do we prepare our students for this challenge? Preparation for life in this globalized world can take multiple forms. Participating in study abroad programs is one way of preparing them to enter the global workplace. But, as constructivist psychologist George Kelly (1963) observed in his groundbreaking book, *A Theory of Personality,* experiences are a matter of how we understand events. As Kelly suggests, we cannot merely thrust students into the intercultural milieu hoping that this will result in their becoming "global citizens." We also need to provide them with educational contexts and opportunities that help them make sense of those events and ultimately to integrate these intercultural experiences into their own cultural identities. We need to prepare them for intercultural interactions before they go, give them tools and strategies for construing these experiences while they are there, and assist them in debriefing and integrating what they have learned upon their return. One of the ways we can do this is by providing them with internationalized courses on their home campuses. Such courses can prepare students for more intense intercultural experiences in living and working abroad, help them to understand how those experiences will change them, and provide ways for them to use international perspectives and experiences within their own cultural contexts.

By making a distinction between experience and perception, Kelly's idea draws attention to the heart of what it means to internationalize on-campus courses/curriculum (IOCC). It explains why IOCC is not merely the addition of a unit on international perspectives or adding a new book introducing intercultural material or case studies, because students are likely to construe or make sense of those materials from their own cultural perspectives and often miss the logic and complexity of the cultural perspective presented.

Further, "ghettoizing" international content to individual units within a given course, or a single course within a larger curriculum continues to set it apart as an "extra" that is not significant enough to be integrated into the students' worldview. As a result, students come to see intercultural communication issues as isolated from, rather than a necessary part of the technical communication tasks they will perform after graduation. Instead, the authors believe that internationalizing is an integrative approach in which intercultural information permeates the entire course, not just a part of it. It requires reconceptualizing the course in a way that includes not just new material, but clear goals, new strategies, and other cultural perspectives. For this reason, the authors believe that successful training programs for internationalizing curriculum are those that seek to shift the objectives of both courses and faculty to be more international in nature.

Ultimately, internationalizing departmental and/or university curriculum in a critical mass of courses is ideal. Research has repeatedly shown that teaching diversity issues in isolated courses does not have a significant impact on the students' attitudes and beliefs (Colville-Hall, MacDonald, & Smolen, 1995; Weisman & Garza, 2002). Students need to see issues of culture and diversity as an integral part of knowing and understanding any body of knowledge and their world.

Kelly's ideas also highlight the rationale for internationalizing courses in technical communication by first internationalizing the faculty. It is the faculty who must ask the hard questions and encourage their students to look beyond their own cultural assumptions and explore other cultural perspectives. For this reason, training programs for IOCC that work with faculty, like the undergraduate and graduate courses these faculty will design, need to be deliberately planned to facilitate transformational change.

This chapter shares the lessons learned about training faculty to teach internationalized courses. It stems from the 3-year *Internationalizing the On-Campus Curriculum* project at the University of Minnesota[1] and proposes both a conceptual framework for internationalizing technical communication courses and specific examples of readings and exercises designed to provide the conditions that can help faculty and students progress systematically through the learning process. Readers are free to adopt these ideas in their own attempts at internationalizing their individual courses or the overall curriculum in which they teach.

UNDERSTANDING TRANSFORMATIONAL LEARNING

The kind of learning that results from internationalizing a course tends to be transformational. This idea is true for both the faculty undergoing training and the students who will ultimately take courses these faculty teach. According to Mezirow (1991), transformational learning

> involves an enhanced level of awareness of one's beliefs and feelings, a critique of their assumptions, an assessment of alternative perspectives, a decision to negate an old perspective in favor of a new one, an ability to take action based on the new one, and a desire to fit the new perspective into the context of one's life. (p. 161)

If one accepts Mezirow's definition, then the participants in an internationalized curriculum, to some degree, need to be "transformed" for the process to be truly effective. This transformation is not a temporary condition. True transformation requires that the participants be changed in ways that significantly affect their worldview and that those changes persist after the transformational experience is over. For example, once one learns how to read (has been transformed from illiterate to literate), it is virtually impossible to not read written postings such as a STOP sign. An apt metaphor would be to imagine a network in which new knowledge interacts and integrates with existing networks of knowledge, organizing, and ultimately transforming it in sometimes surprising and unanticipated ways.

TRANSFORMATIONAL LEARNING AND THE INTERNATIONALIZED COURSE

O'Donovan and Mikelonis (2005) identified four basic assumptions to keep in mind when internationalizing a course:

[1] The Internationalizing the Curriculum Project was sponsored by a Bush Foundation grant awarded to the Office of International Projects in cooperation with the Center for Teaching and Learning Services at the University of Minnesota from 2000 to 2004.

> **Assumption 1:** Internationalizing is an intentional approach to constructing new knowledge and designing courses.
>
> **Assumption 2:** The process of international/intercultural education is provocative by its very nature because it challenges deep-seated attitudes, beliefs, and values.
>
> **Assumption 3:** Pedagogical practices in international courses promote self-discovery, self-reflection, and personal transformation.
>
> **Assumption 4:** Typical obstacles to internationalizing curriculum include resistance to change, lack of tolerance for ambiguity, and the inability to reflect critically.

These assumptions raise important ethical concerns as well. When we put students into situations in which they will confront their tacit assumptions and strongly held cultural beliefs, values, and behaviors, we are pulling the "cultural rug from under their feet," and need to be sure we are able to catch them before they "fall." Since the essential ethical message here is: *Do no harm* (Smith, 1993), this is a task that should not be undertaken lightly. Intercultural educators need to assume responsibility for helping students to coherently reconstruct the cultural realities that they may be dismantling (Smith, 1998). In so doing, the instructor needs to safely and effectively guide students through the all-important transformational learning experiences needed for effective learning related to culture and communication.

Because it is not enough for instructors to challenges students' cultural realities, we have added three additional principles to the previous four. Faculty must also be responsible for

> **Assumption 5:** Assisting their students in their struggle with reorienting their assumptions about issues of rightness and wrongness.
>
> **Assumption 6:** Helping guide students in understanding that context and cultural realities must be considered before information and behaviors can be evaluated. (Bennett, 1993)
>
> **Assumption 7:** Helping students to develop an awareness of, and be mindful about, their ability to make choices in their development of an overriding ethical structure that can guide them in their own journey toward intercultural awareness, i.e., a view in which "ethnorelativism and a strong ethical principles coexist. (Bennett, 1993, p. 30)

Because the authors believe that global interdependence is a reality with which our technical communication students will need to deal both professionally and personally (Friedman, 2005), these ethical concerns become risks worth taking as we prepare students to function effectively and compassionately on a multicultural playing field.

The following sections of this chapter detail the authors' own experiences in taking such an approach to teaching students about intercultural technical communication.

SELECTING MODELS FOR INTERNATIONALIZING COURSES

The challenges for instructors of internationalized courses are multiple and involve five key questions:

1. How can we induce these changes in learners?
2. How can we plan the change?
3. How can we sequence activities and readings designed to lead students toward this transformation?
4. How will we know if the transformation has taken place?
5. How can we measure that change?

Finding the answers to these questions is at the heart of an internationalized curriculum. During a 3-year project to internationalize the curriculum at the University of Minnesota,[2] two important theoretical models guided the sequencing and presentation of information and tasks, and the strategies employed. First was Bennett's (1993) *Developmental Model of Intercultural Sensitivity* (DMIS), which presents a framework for understanding the stages a person goes through when encountering cultural differences (see Figure 1, Bennett's Developmental Model of Intercultural Sensitivity). The underlying assumption of the model is that, as one is able to think in more complex ways about intercultural interactions and experiences, one moves through stages of *ethnocentrism* and ultimately experiences a paradigm shift into stages of *ethno-relativism,* leading to increasingly sophisticated understandings and behaviors around the issue of cultural difference.

The second theory was a *transformational model* for internationalizing courses (see O'Donovan, Mikelonis, & Smith, 2005) and consists of three stages:

1. Generating Consciousness
2. Transforming Consciousness
3. Expanding Consciousness

[2] See footnote 1.

Ethnocentric Stages

I. Denial: No consideration that difference exists
   A. Isolation: No exposure to cultural differences; parochial or stereotypical categories
   B. Separation: Physical or social barriers keep "others" at a distance

II. Defense: Difference is threatening; the defense is psychological
   A. Denigration: Difference is considered "bad" or "wrong"
   B. Superiority: They're okay, but we're better
   C. Reversal: They are "right" and we are "bad"

III. Minimization: People are people and the differences are superficial
   A. Physical Universalism: We're all basically "human"
   B. Transcendent Universal: All people really need to believe a certain way to be happy

Ethnorelative Stages

IV. Acceptance: Difference is both acknowledged and respected
   A. Respect for Behavioral Differences
   B. Respect for Value Differences

V. Adaptation: Skills for relating to and communicating with other cultures are enhanced
   A. Empathy: Ability to imagine and understand the other's reality
   B. Pluralism: Commitment to a philosophical frame that includes multiple cultural views

VI. Integration: Integration of multiple cultural views into one's psychological repertoire
   A. Contextual Evaluation: Ability to make judgments in a variety of cultural contexts
   B. Constructive Marginality: Integration of multiple cultural views into one's identity

Figure 1. Bennett's Developmental Model of Intercultural Sensitivity (DMIS).

These theoretical models guided the internationalized curriculum development, first in the authors' work with the faculty who volunteered to internationalize their courses and later, when faculty used variations of these same models to the internationalize the courses they designed for their undergraduate students. Additionally, the authors found the *consciousness competence learning model* (Howell & Fleishman, 1982) to be an effective explanatory addition to the mix (see Figure 2, Conscious Competence Learning Model), particularly with regard to the value of using the *Intercultural Development Inventory* (Bennett & Hammer, 1999) as a tool to help faculty become more aware of their assumptions around difference.

All of the models will be discussed in more depth within the context of how they were applied in the training.

APPLYING THEORY TO INTERNATIONALIZED COURSES

The next section of this chapter examines the implementation of these models in the faculty training and shares some of the exercises and readings that were used to assist faculty in developing their internationalized courses. The faculty adapted many of these readings and exercises for use with their own students. The term *facilitator* is used to describe the role of the teacher/trainer in the workshops, because the kind of learning occurring in such an environment should be generative, and meaning needs to be co-created between and among facilitators and participants.

Stage 1: Generating Consciousness

Generating consciousness is the first stage in the process of internationalizing, and it involves helping participants to recognize their own ethnocentrism by making them aware of their own embodied knowledge, tacit assumptions, and perspectives. In this case, ethnocentrism[3] is defined *perceptually* as a lack of awareness of other cultural perspectives. We have found that "de-demonizing" ethnocentrism helps to facilitate openness to challenging what we know and accepting new knowledge and perspectives. Thus, this stage of the process is designed to move participants from *unconscious incompetence* to *conscious incompetence* with regard to recognizing difference and teaching about culture (see Figure 3, A Transformational Model for Development of Internationalized On-Campus Courses). Helping faculty, whose identities are to some degree based on their role as "expert," recognize where their blind spots are; that is, recognizing what they don't know can be a somewhat daunting task.

[3] It should be made clear here that ethnocentrism is not viewed as fundamentally bad; in fact, it plays a key role in the creation and maintenance of cultural systems and allows people within those systems to function in an environment that minimizes ambiguity and maximizes the potential for understanding. In intercultural interactions, however, it can also be the chief cause of misunderstanding and conflict.

Competence

Conscious 3 - Conscious Competence

- need to concentrate and think in order to perform the skill
- can perform the skill without assistance
- thought is required for the task to be reliably performed: it is not yet second nature
- can demonstrate the skill to another, but is unlikely to be able to teach it well to another person
- continues to practice the new skill, and, if appropriate, commit to becoming "unconsciously competent" at the new skill
- *practice* is the single most effective way to move from stage 3 to 4

Unconscious 4 - Unconscious Competence

- the skill enters the unconscious parts of the brain—it becomes second nature
- it becomes possible for certain skills to be performed while doing something else
- ability to teach others the skill, although after some time of being unconsciously competent, the person might actually have difficulty in explaining exactly how they do it: the skill has become largely instinctual
- longstanding unconscious competence needs to be checked periodically against new standards

Incompetence

2 - Conscious Incompetence

- aware of the existence and relevance of the skill
- aware of a deficiency in this area, ideally by attempting to use the skill
- realizes the need to improve the skill or ability in order to increase effectiveness
- understands what level of skill is required for his/her own competence
- makes a commitment to learn and practice the new skill, and to move to the "conscious competence" stage

1 - Unconscious Incompetence

- unaware of the existence or relevance of the skill area
- unaware of a particular deficiency in the area concerned
- might deny the relevance or usefulness of the new skill
- must become conscious of their incompetence before learning can begin
- the learner must move into the "conscious incompetence" stage by becoming aware of the skill or ability and the benefit that it will bring to the person's effectiveness

Figure 2. Conscious Competence Learning Model.

STAGE I: Generating Consciousness (Unconscious Incompetence → Conscious Incompetence)

OUTCOME: An increased awareness of tacit assumptions that influence personal frames or worldviews.

TASKS

- Create an environment of trust
- Establish a shared purpose and language
- Surface embodied knowledge

INSTRUCTIONAL STRATEGIES (experiential in nature)

- Case Studies, Simulations, Role-plays
- Shock and Aha!

ISSUES

- Resistance
- Fear, risk of appearing ignorant or politically incorrect

STAGE II: Transforming Consciousness (Conscious Incompetence → Conscious Competence)

OUTCOME: An ability to question existing assumptions and to tolerate ambiguity

TASKS

- Compare and contrast original assumptions
- Question and deconstruct original assumptions
- Negotiate factors that create resistance

INSTRUCTIONAL STRATEGIES

- Case studies
- Debates
- Constructed controversies
- Attendance to student cognitive development, ability to shift frame of reference and seek emic meaning from readings and experiences, resistance, and discomfort due to challenging existing worldviews

ISSUES

- Resistance and discomfort due to challenge to existing worldviews

STAGE III: Expanding Consciousness (Conscious Competence → Unconscious Competence)

OUTCOME: An expanded set of personal assumptions that inform the design, development, and presentation of an internationalized course

TASKS

- Reconstruct new knowledge, attitudes and behaviors
- Integrate new knowledge, attitudes, behaviors into course design, development, and presentation

INSTRUCTIONAL STRATEGIES

- International speakers and resources
- Discussion, simulations, discovery, and exploration of new perspectives

ISSUES

- Learning how to learn
- Continue to engage with cultural differences

Figure 3. A Transformational Model for Development of Internationalized On-Campus Courses.

To this end, all participants were given the IDI (Bennett & Hammer, 1999) prior to the first meeting. Based on Bennett's *Developmental Model of Intercultural Sensitivity*, the IDI is a 50-question paper/pencil test designed to identify the stages of intercultural development a person has "resolved"; that is, how they have come to grips on a personal and cognitive level with the concept of difference. A discussion of the Group Profile and the theory on which it is based (Bennett's DMIS) was carried out with the group as a whole. Participants were also given their individual results and, during an individual consultation, discussed the implications of those results and developed a plan of action.[4] One of the key advantages of the instrument as a counseling tool is that it allows for measuring both where people would like to be and where they actually are. This was always seen as a gentle process in which we collaborated with participants in setting goals for their own development.

The decision to be prescriptive with regard to our faculty's development toward achieving an ethnorelative perspective on the DMIS was not made lightly. The authors strongly believe that anyone who chooses to teach "other people's children" about intercultural issues needs to be aware of the assumptions and values they personally bring to the classroom. To do this from an ethnocentric posture would be both unethical and, to some degree, dangerous.

Making progress in internationalizing requires developing a shared purpose and language (e.g., How are culture, ethnocentrism, ethnorelativism, etc. defined?) and creating a climate of trust, where participants feel safe in taking risks and exposing their own vulnerabilities. Experiential learning strategies using case studies, simulations, and role-plays that induce awareness of the implicit cultural assumptions that we all hold are vital at this stage of the process. Throughout this process, intercultural educators need to be prepared to deal with participant resistance to the process, a resistance based primarily on fear of appearing foolish, ignorant, or politically incorrect.

In an effort to generate awareness of how automatic and embedded our assumptions are about the way things "ought to be," workshop facilitators quickly moved participants out of their comfort zones by providing them with experiences that were foreign to them, using a technique referred to as "shock and aha!"[5] At the beginning of the first session, workshop facilitators engaged participants in an activity designed to surprise them or fly in the face of their expectations. For example, at the first daylong retreat with faculty, facilitators greeted them at the door in stocking feet and wearing colorful shawls, with smiles and

[4] Giving the IDI requires certification that is obtained by taking a 3-day training course. One of our trainers is certified, but in most cities it is relatively easy to find someone who is IDI certified and can oversee this process for you.

[5] "Shock and Aha!" is a name that Kathleen O'Donovan, one of the facilitators, gave these activities, as a take off on the "shock and awe" tactics of the American military in the Iraq War.

welcoming gestures. Soft atonal instrumental music was playing in the background; unfamiliar proverbs and pithy sayings appeared on the overhead screen without explanation or commentary.

Facilitators communicated nonverbally throughout this period. We indicated that the participants were to take off their shoes before entering the training room, and then one of us graciously led them to their seats, which were arranged in a semicircle. When they were all assembled, we joined the group and motioned for them to rise, leading them in a Tai Chi exercise. All of this was done without saying a word. After some initial tittering, the participants followed our lead and seemed willing to do what they were asked, even though this was not typical of their expectations for a training session. One participant was particularly irritated, refusing to take off her shoes because the "floor in the training room is cold," and finally left the room for a short time to "get a pair of socks." She did, however, eventually return and rejoined the group.

The process continued in silence for about 15 minutes. Then facilitators debriefed with participants on what their tacit assumptions were about, how to welcome people to a training session, their expectations about the dignity inherent in their roles as faculty, how the exercise frustrated them, how they felt about what was going on, and ultimately, why they decided to play along.

This was the first of many activities designed to generate participant awareness that something was going on that they did not quite understand; but it was something that required both their participation and their suspension of judgment. Initially, they granted us their trust and good naturedly participated in the activities the facilitators had planned. Later, however, they would find themselves more challenged by the activities and more frustrated by their reactions to them. The participants would also find themselves more irritated because of what the facilitators expected of them.

The first daylong session was dedicated to creating an environment of trust. Workshop facilitators shared their own intercultural experiences and led participants through various exercises designed to elicit their trust in the facilitators, the process, and in one another. The activities and exercises helped to create a "community of learners" and to establish a shared purpose and language, which would form the basis for our discussions throughout the other sessions (see the *Internationalizing the Curriculum Workshop* Syllabus.)

D.I.E. EXERCISES TO RAISE AWARENESS OF CULTURE AND COMMUNICATION

Early in the training, participants were introduced to the D.I.E. (Describe, Interpret, Evaluate),[6] an exercise designed to teach participants a method for

[6] Bennett, Bennett, & Stillings (1977).

analyzing unfamiliar experiences and artifacts without rushing to judgment, because we wanted the participants to begin to use it throughout the course. The point of the exercise was to get participants to learn, when faced with an unknown event or ambiguous information, to first *describe*, as objectively, neutrally, and accurately as possible, what they see; *interpret* what they see based on that description; and *evaluate* (good/bad, beautiful/ugly, etc.) based on the interpretation(s) generated. Any ambiguous object can be used for this exercise. In this case, it was a largely unidentifiable combination of sticks, twine, and shells from Micronesia. Participants were told to, "Say something about this—whatever pops into your head." Participants' responses were written under one of three columns on the board headed by the letters D, I, and E, depending on whether they were descriptions, interpretations, or evaluations, making sure that everyone responded at least once. In the first round, participants' responses generally weighed most heavily toward interpretations (which are most often culture/experience bound), because people tend to try to make sense of something unknown by using their existing conceptual frameworks. In the second round, they were asked to concentrate first on describing the item and then pressed to interpret it based on the descriptions previously given. Generally someone would eventually (closely) determine what the object was: in this case, a map of water currents in the Trukese Archipelago.

Workshop facilitators also provided participants with opportunities to identify their implicit values, assumptions, and mental models and compare them with different cultural perspectives through the use of stories, case studies, exercises, and participant discussion and debate around key issues (i.e., cultural values, behaviors, and culturally embedded ethical concerns). These activities often elicited a "shock and aha!" response from the participants. The "shock" came from the clash in cultural values, attitudes, and beliefs that led to behaviors that seemed inappropriate from an American perspective. The "aha!" came from understanding (not necessarily agreeing with, but understanding) the underlying values, attitudes, and beliefs that resulted in the behaviors exhibited by the members of a different culture.

Our goal in leading participants through these activities was to help them learn how to suspend judgment (at least until they had learned a great deal more about the culture in question) when faced with intercultural situations that they did not understand. Because such judgments sometimes carry an emotional, even gut-wrenching component, this is easier said than done and requires practice. Understanding how such behaviors are situated within a cultural context allows for a more accurate understanding of the purpose of those behaviors and ultimately the ability to make more astute judgments about their usefulness and effectiveness.

Stage 2: Transforming Consciousness

During this second stage of the internationalizing process, the goal is to push participants to undergo a major paradigm shift from ethnocentrism to ethnorelativism; to increase their ability to question, and even to challenge existing assumptions; and to become more comfortable with the ambiguity inherent in acknowledging that there is more than one "right" way for people and cultures to be. In this process, participants move from *conscious incompetence* to *conscious competence* in their ability to apply new strategies and concepts to their course development and to anticipate and negotiate issues that create resistance to alternative cultural perspectives.

At this stage, case studies become more challenging, debates more invested, and controversies more intense. Examples are specifically chosen to jolt them out of their universal "we are all basically the same" position. Participants are given information about models that classify cultural dimensions so that they can use them prototypically (not stereotypically) to form hypotheses about different cultural behaviors. Special attention is paid to cognitive development, the ability to shift frames of reference, and practice in seeking *emic* (culture specific) meaning from readings and experiences. At this point, intercultural educators need to take great care in supporting participants as they confront repeated challenges to their existing worldviews. Failure to navigate this stage successfully can result in retreat and entrenchment in ethnocentrism and its more comfortable assumptions about the predictability of human behavior.

One case study proved particularly successful in producing visceral reactions and surfacing cultural differences. The case involved the true story of a Japanese woman living in California who drowned her two children before attempting her own suicide when she discovered that her husband had a mistress. Both children died, but a passerby rescued the mother, who was then charged with murder. Because the events took place while she and her husband were living in the United States, the case became a major news story (Reese, 1985) partly because of the different cultural assumptions and norms surrounding the event.

Participants were horrified by the woman's actions and puzzled by the husband's lack of recrimination of her behavior. As participants reacted to and attempted to analyze the story using the D.I.E. exercise, the facilitators continued to provide them with information about the cultural dimensions, including some of the basic characteristics of high- and low-context cultures, and specific Japanese values, attitudes, and beliefs that had a direct bearing on this situation. The group also discussed the effects of cultural adjustment on people immersed in (and in this case, isolated in) a foreign culture. Although the participants did not agree with the Japanese assumptions about the situation and the mother's culpability, most were able to understand them from a Japanese perspective.

To further challenge participants to internalize the lessons of the D.I.E. exercise, suspend judgment, and increase their tolerance for ambiguity, facilitators presented a cross-cultural simulation called "The Albatross." The simulation provides an opportunity for participants to experience the process of observing and participating in another culture and understanding how we often (mistakenly) judge the behavior of others based on our own cultural assumptions. During the simulation, participants observe certain ritualistic behaviors by a man and a woman who are members of the Albatross culture. Participants are then asked to describe the culture, to make interpretations, and to draw conclusions about the cultural values and practices of the Albatross people.

When asked to describe the Albatross culture, typically, about one-fourth to one-third of the participants were able to describe what they had seen but refrained from interpreting or evaluating the actions of the actors. Others, however, immediately slipped into interpretations based on their own cultural lenses. They described the culture as patriarchal (the women as subservient to the men) and the culture as quite primitive. As the facilitators guided the discussion and ultimately explained the culture, it became clear that the underlying cultural assumptions of the Albatross people were entirely different from the interpretations initially made by the observers. The facilitators demonstrated how the actions can be interpreted in two totally different and diametrically opposed ways depending on the underlying cultural assumptions and values they apply. Ultimately participants realized that assumptions of similarity,[7] when applied to another culture, can hamper the understanding of cultural difference. An additional component of this analysis can be questioning the assumption that issues of power and dominance always figure into the way we view and evaluate the behavior of others and ourselves.

At this stage in the transformational process, participants have confronted numerous discrepancies between their own cultural perspectives and those of other cultures. It becomes increasingly difficult for them to ignore or keep from questioning the universality of their cultural values and beliefs. They begin to realize that what they consider "reality" or "the way things are" is to some degree culture-bound and varies from culture to culture. The task at this phase is to take this new awareness of "other ways of being in the world" and integrate it into their worldviews.

[7] When we observe actions or behaviors in another culture that are similar to actions or behaviors in our own culture in similar circumstances, we often impute motives and interpret those actions based on our own cultural assumptions and values. Whenever we do this, we run the risk of being partially or even completely wrong—both in our interpretations of the actions and in our imputation of motives for those actions. We have to learn enough about the other culture to understand their values, beliefs, and attitudes before we can understand the motives for their behaviors and how they fit into the complexity of a given cultural context.

When the participants reach this stage, they are challenged with expanding and changing their sense of *who they are*. This can be an uncomfortable and even painful process. But it is a necessary one if participants are to move along the continuum from ethnocentrism to ethnorelativism, because it involves a paradigm shift in their understanding and behavior. This obvious struggle indicates that deep learning is taking place. This awareness can produce enough stress and cognitive dissonance that some participants will lash out at the facilitators rather than reconcile what they are experiencing with their own deep-seated cultural values. This marks a milestone in the training, one that requires facilitators to be compassionate, supportive, and to proceed carefully and skillfully to assist participants in the process of self-discovery, self-reflection, and personal transformation.

Stage 3: Expanding Consciousness

In the third stage of the internationalizing model, participants continue to undergo significant change and develop expanding awareness and skill in dealing with cultural difference. The tasks now are to reconstruct and integrate new knowledge, attitudes, and behaviors into their personal and professional worldviews. At this stage, participants are settling into a more ethnorelative perspective, and are internalizing constructs that make them more open to recognizing, confronting, and embracing difference. Participants have learned "how to learn" in an intercultural environment, and they continue to engage with cultural difference. However, because cultural sensitivity and culture learning are ongoing processes that provide, according to Janet Bennett, "an infinite opportunity to make a complete ass of yourself in every culture in the world,"[8] it is vital that participants are able to reflect on their experience and translate their experiences into an awareness of how culture learning takes place. At this stage, participants have had sufficient encounters with cultural difference that they begin to automatically adapt and adjust, using strategies that enable them to explore cultural differences without using their own culture as the only touchstone for their experience. It is this awareness that reflects the movement from *conscious competence* to *unconscious competence*, and makes it possible to continue to engage and learn about cultural difference while applying that knowledge in the design, development, and presentation of their courses.

This is often an optimistic, heady, and productive time. Course planning takes on a new dimension because participants look at their content with new eyes and are often excited about sharing their insights with others who have also gone though this process. They are often nervous about how their course will be received by their peers, but they look forward to discussing their syllabi, and

[8] J. M. Bennett, personal communication, 1985.

brainstorm innovative ways of presenting the content of their courses with others in the class and the facilitators, develop multimedia presentations, and eagerly explore different perspectives. They created "shock and aha!" exercises and seek out artifacts, pictures, and events to be used as the content for D.I.E. activities. At this stage, participants learn to make strategic use of silence and reflection in their teaching. They also seek out others who have, or are developing, similar intercultural perspectives and use them as sounding boards for continuing to progress along the ethnorelativist continuum.

ORGANIZATION AND OBJECTIVES IN INTERNATIONALIZING LEARNING

The first three stages of the internationalizing model are not linear, but rather they build on one another in a fluid, integrative way. Different participants progress at different rates through these stages, so intercultural facilitators must be closely engaged with the participants in order to gauge the rate of change occurring and to lend support and assistance when needed. At the third stage, participants become aware of new and different cognitive models, and they will continue to progress if they can communicate with, and gain support from, others who are going through the same process. In this way, they coalesce into a community of learners who are engaged and interdependent. By shifting to an expanded view of reality and committing to new perspectives on teaching and learning, the participants are able to design courses that are both intellectually challenging and emotionally engaging.

Stage 4: Wrapping it Up and Sending Them Off

To demonstrate what they had learned, participants prepared an annotated syllabus of their internationalized course that gives details of goals and objectives; content, process and evaluation; and provides rationales for key decisions made in the course's development. The course syllabus demonstrated how the course provided students with (a) an increased understanding of their own cultural perspectives and how those perspectives color the students' view of reality and (b) an awareness of diverse perspectives regarding the course content.

To assist them in this process, the facilitators challenged the faculty to question their own assumptions about the way a course is designed, delivered, and evaluated, as well as how content can be chosen, integrated, and sequenced into their proposed courses. The facilitators provided opportunities to learn about and grapple with models of objectives-based (or "backward") course design (Wiggens & McTighe, 1998). The facilitators also modeled teaching strategies (recounting those used in the training program) that encourage self-reflection, active learning, and personal transformation. Finally, participants were challenged to change their perceptions of their role of "instructor" from purveyors of knowledge to facilitators of learning, from classroom manager to learning guides who create a climate where students can grapple honestly with their reactions to new ideas and perspectives, and finally from taking responsibility for information to taking ethical responsibility for their student's emotional and behavioral learning. These activities and discussions were designed to assist participants in reconceptualizing their course content to integrate international perspectives, identify performance objectives, and engage in formative assessments and assignments that will demonstrate how well the students have learned what was intended.

SUMMARY AND CONCLUSIONS

People do not generally embrace change easily. When our identities are threatened because of the need to integrate different, often contradictory, cultural attitudes, beliefs, and behaviors, we can either retreat into ethnocentrism or embrace the challenges and grow from the experience. If we embrace change, we must also realize that we may never be able to view our own culture, or the world in general, in quite the same way again. The authors believe that embracing the change that results from allowing other cultural perspectives to elaborate our worldview can increase our ability to understand cultures different from our own and encourage us to continue to assimilate culture learning tools that will enable us to continue to learn and grow. The authors viewed the internationalizing workshops not as an end in themselves, but as the beginning of each participant's personal journey on an intercultural path.

The systematic process of leading faculty and students through the stages of generating consciousness, transforming consciousness, and expanding consciousness proved an effective and theoretically sound way of reconceptualizing course goals and content, and thus led faculty on a path to successfully internationalize undergraduate courses. In so doing, established intercultural education principles and sound pedagogical practice were combined with the content of a variety of different disciplines to provide students with a more global perspective in their undergraduate education. The authors believe it can help both faculty and students move along the continuum from ethnocentrism toward ethnorelativism, thereby broadening their worldviews and expanding their sense of self and if we're lucky, nudge us all one step closer to peace and understanding.

A final note: Evaluations and focus groups of students and faculty following the actual presentation of the internationalized courses showed the courses to be remarkably effective in increasing student interest in both the course material and things international. Many said that they had never considered international issues or the possibility of studying abroad before, but now were interested in further exploring both.

APPENDIX
Internationalizing the Curriculum Workshop Syllabus
Bush Grant 2002–2004

Internationalizing Curriculum Goals and Outcomes

Workshop

Design and implement a comprehensive faculty development model to internationalize undergraduate curricula.

1. Foster the development of multidisciplinary networks among faculty interested in transforming the teaching and learning that occurs in their classes.
2. Encourage and support a university culture that acknowledges the importance of internationalizing curricula.
3. Create a venue for faculty to discuss internationalizing one of their courses.
4. Develop new pilot efforts designed to contextualize the study-abroad experience through the internationalization of on-campus courses.
5. Encourage faculty to question implicit and explicit assumptions surrounding international and cross-cultural perspectives.
6. Develop and articulate a process to design, develop, present, and evaluate an internationalized course.
7. Support participants in their production of an internationalized syllabus.
8. Help faculty incorporate effective pedagogical strategies in their efforts to internationalize curriculum.
9. Support participants in their production of an internationalized syllabus.
10. Help faculty incorporate effective pedagogical strategies in their efforts to internationalize curriculum.

Faculty Outcomes

Faculty will be able to

1. Help prepare students to perform more effectively in a global context through increased international and cross-cultural competence.
2. Integrate an international perspective into their undergraduate courses.
3. Encourage their students to question implicit and explicit assumptions surrounding international and cross-cultural perspectives.
4. Design, develop, present, and evaluate an internationalized course.
5. Create an internationalized syllabus.
6. Incorporate effective pedagogical strategies in their efforts to internationalize curriculum.

Basic Tenets

- Internationalizing the curriculum involves incorporating global perspectives and content in all courses—it does not mean "ghettoizing" international issues in a single course.
- Internationalizing the curriculum can involve change on multiple levels by
  1. Including international content in the curriculum;
  2. Using international faculty and students as resources;
  3. Providing for and drawing on faculty experiences abroad (research, site visits, Peace Corps, etc.);
  4. Using international centers on campus (Minnesota International Center, the China Center, Global Campus, Office of International Programs, International Student and Scholar Services) as resources;
  5. Integrating content learning abroad with second-language learning;
  6. Integrating content studies abroad with content in the United States;
  7. Valuing and encouraging U.S. students who have studied abroad (i.e., How can curriculum be designed to help them use that knowledge and experience?);
  8. Valuing and encouraging international students' experience and perspectives (i.e., How can curriculum be designed to help them integrate course knowledge and experience into their home country context?);
  9. Developing courses and capstone experiences abroad by using existing resource sites and programs with the assistance of Global Campus and the Office of International Programs;
  10. Developing and incorporating student-centered pedagogical strategies that teach to and reward multiple learning and cultural styles;
  11. Using technology (Internet, listserv, etc.) to provide forums for dialogue between students, classes, institutions, and nations;
  12. Developing opportunities for independent study and research with the assistance of the Global Campus and Office of International Programs and/or through existing programs such as SPAN (Student Project for Amity Among Nations) & MSID (Minnesota Studies in International Development).

Our Philosophy: Internationalizing a course involves the recognition that a fully integrated curriculum will affect students and faculty on a number of levels. Instructors must be prepared not just to provide international content, but also to be able to facilitate students as they grow and develop a more international perspective. This is not an easy journey. At some points, it is inevitable that students will experience culture-bound resistance during their interpretation and evaluation of the information they are receiving. As a result, faculty must be prepared to engage with them intellectually, psychologically, and emotionally in the process.

This workshop will provide a learning community where we can work together to develop a quality curriculum, anticipate potential pitfalls, and prepare for the process of guiding students through the challenges of an internationalized course. This will involve a degree of personal reflection and growth as you work through the process.

Schedule

Session 1 ***Becoming Familiar with Internationalizing the Curriculum (Retreat)*** October 3

9:00–9:45	Welcome, Introductions, Overview
9:45–11:00	IDI Results and Discussion
11:00–12:00	Cultural Dimensions
12:00–1:00	Lunch and Expectations of International Programs
1:00–2:00	Metaphors and Assigned Reading Schematic Model of Intercultural Communication
2:00–3:00	Shifting Perspectives and Assigned Reading "Domains of Influence for Curriculum Design"
3:00–3:45	Program Assumptions
3:45–4:00	Wrap-up and Round Robin

Homework for October 24

1. Focus on three areas from the Domains of Curricular Influence model.
2. Identify your assumptions concerning each domain if the course were to be internationalized and not internationalized.
3. Generate a preliminary outline for your intended course.
4. Focus your attention on only two outcomes.

Session 2 ***Developing an Internationalized Course*** October 24

9:15–9:45	Participants' Intentional Exercise: Noticing, Breaking Patterns, Questioning the Received Wisdom
9:45–10:30	Homework from the Retreat: Domains of Influence (Carousel Activity)
10:30–11:15	Discussion of assigned readings for October 24: Jigsaw #1 Barr and Tagg. "Comparing Educational Paradigms" (everyone) Freedman, Kerry. "Culture in Curriculum" Mestenhauser, Joseph A. "Portraits of an International Curriculum" Bennett, Milton J. "Towards Ethnorelativism"

Homework for November 7

1. Draft the syllabus for your internationalized course.
2. Review the syllabus that appears in your reading packet. Answer the following questions:
   - What are the positive features of this syllabus?
   - How might it be improved?
   - What strategies to internationalize did he incorporate?

Session 3 ***Creating an Internationalized Syllabus*** November 7

9:15–9:45	Participants' Intentional Exercise: Noticing, Breaking Patterns, Questioning the Received Wisdom
9:45–10:15	Homework from Session 2: Draft of Syllabus (pairs exercise)
10:15–11:00	Bloom's Taxonomy: What's missing? Turning objectives into outcomes The difference between assessment and evaluation
11:00–11:20	Discussion of Dobbert article and the sample syllabus

Assigned readings for November 7th:

Collins, Terence. "For Openers . . . An Inclusive Course Syllabus"
Dobbert and Marion. "The Impossibility of Internationalizing Students by Adding Materials to Courses"
Handout on Assessment/Evaluation
Considerations Packet for Course and Syllabus Design

Homework for November 21

1. Revise your syllabus.
2. Read the article assigned to you for Jigsaw #2.
3. Bring in your definition of culture shock.

Session 4 ***Relating Change to Transformational Learning*** November 21

9:15–9:45	Participants' Intentional Exercise: Noticing, Breaking Patterns, Questioning the Received Wisdom
9:45–10:15	SGID: An evaluation strategy
10:15–11:00	Jigsaw # 2: Mikelonis, Victoria "The Message Sent vs. the Message Received" Mezirow, Jack. "Perspective Transformation: How Learning Leads to Change" Brookfield, Stephen. "What It Means to Be a Critically Reflective Teacher"
11:00–11:20	Perspectives on Culture Shock

Homework for December 4

1. Revise your syllabus and be prepared to discuss it.
2. Read the article assigned to you for Session 5.
3. Fill out the posttest instrument.

Session 5 ***Identifying and Responding to Challenges in Internationalizing Curriculum*** December 4

9:15–9:45 Participants' Intentional Exercise: Noticing, Breaking Patterns, Questioning the Received Wisdom

9:45–10:30 Homework: Pairs discussion of revised syllabus

10:30–11:15 Jigsaw # 3:

Paige, R. Michael. "On the Nature of Intercultural Experiences and Intercultural Education"

_____. "Trainer Competencies for International and Intercultural Programs"

Paige and Martin. "Ethics in Intercultural Training"

Smith, Shelley, "Ethics in Intercultural and Diversity Training"

Homework for February

Prepare the final version of your syllabus and be prepared to present it in February 2004 at our final meeting.

References

Bennett, Milton J. (1993). Towards ethnorelativism: A developmental model of intercultural sensitivity, In R. M. Paige (Ed.), *Education for the intercultural experience* (pp. 21–71). Yarmouth, ME: Intercultural Press, Inc.

Brookfield, Stephen D. (1995). *Becoming a critically reflective teacher* (pp. 1–27). San Francisco: Jossey-Bass.

Clark, M. Carolyn (1993, Spring). Transformational learning. In *New directions for adult and continuing education, 57*(1), 47–56.

Collins, Terry (1997). For openers: An inclusive course syllabus. In W. E. Campbell & K. A Smith (Eds.), *New paradigms for college teaching* (pp. 79–102). Edina, MN: Interaction Book Co.

Dobbert, L., & Marion, L. (1998). The impossibility of internationalizing students by adding materials to courses. In J. A. Mestenhauser & B. J. Ellingboe (Eds.), *Reforming the higher education curriculum: Internationalizing the campus* (pp. 53–67). American Council on Education: Oryx Press.

Freedman, Kerry (1998). Culture in curriculum: Internationalizing learning by desire. In J. A. Mestenhauser & B. J. Ellingboe (Eds.), *Reforming the higher education curriculum: Internationalizing the campus* (pp. 40–53). Phoenix, AZ: American Council on Education and Oryx Press.

Gannon, Martin J. (2000). American football. In *Understanding global cultures: Metaphorical journeys through 23 nations* (2nd ed., pp. 209–226). Thousand Oaks, CA: Sage.

Johnston, J. S., & Edelstein, R. J. (1993). Beyond borders. In J. S. Johnston Jr. & R. J. Edelstein (Eds.), *Beyond borders: Profiles in international education* (pp. 1–23). Washington, DC: AAC&U.

Mestenhauser, Josef A. (1998). Portraits of an international curriculum: An uncommon multidimensional perspective. In J. A. Mestenhauser & B. J. Ellingboe (Eds.), *Reforming the higher education curriculum: Internationalizing the campus* (pp. 3–39). Phoenix, AZ: American Council on Education and Oryx Press.

Mezirow, Jack. (1991). *Transformative dimensions of adult learning* (pp. 145–195). San Francisco: Jossey-Bass.

Mikelonis, Victoria M. (2000). Message sent versus message received: Implications for designing training materials for central and eastern Europeans. In P. J. Hager & H. J. Scheiber (Eds.), *Managing global communication in science and technology* (pp. 207–231). New York: John Wiley & Sons, Inc.

Paige, R. M. (1993). On the nature of intercultural experiences and intercultural education. In R. M. Paige (Ed.), *Education for the intercultural experience* (pp. 1–19). Yarmouth, ME: Intercultural Press, Inc.

Paige, R. M. (1993). Trainer competencies for international and intercultural programs. In R. M. Paige (Ed.). *Education for the intercultural experience* (pp. 169–193). Yarmouth, ME: Intercultural Press, Inc.

Paige, R. M., & Martin, J. M. (1996). Ethics in intercultural training. In D. L. Baghat & R. S. Baghat (Eds.), *Handbook of intercultural training* (2nd ed., pp. 148–164). Thousand Oaks, CA: Sage.

Smith, S. L. (1998). Ethics in intercultural and diversity training: A guiding perspective. *Journal of the Northwest Communication Association, 26*(2), 25–39.

REFERENCES

Bennett, M. J. (1993). Towards ethnorelativism: A developmental model of intercultural sensitivity. In R. M. Paige (Ed.), *Education for the intercultural experience* (pp. 21–71). Yarmouth, ME: Intercultural Press, Inc.

Bennett, M. J., Bennett, J. M., & Stillings, K. (1977). *ICW workshop facilitator's manual.* Portland, OR: Portland State University.

Bennett, M. J., & Hammer, M. (1999). The intercultural development inventory (IDI). Portland, OR: Intercultural Communication Institute.

Colville-Hall, S., MacDonald, S., & Smolen, L. (1995). Preparing pre-service teachers for diversity in learners. *Journal of Teacher Education, 46*(4), 295–303.

Friedman, T. (2005). *The world is flat*. New York: Farrar, Straus & Giroux.

Howell, W. C., & Fleishman, E. A. (Eds.). (1982). *Human performance and productivity,* (Vol. 2): *Information processing and decision-making*. Hillsdale, NJ: Erlbaum.

Kelly, G. A. (1963). *A theory of personality*. New York: W. Norton & Co., Inc.

Mezirow, J. (1991). *Transformative dimensions of adult learning* (pp. 145–195). San Francisco: Jossey-Bass.

O'Donovan, K., Mikelonis, V. M., & Smith, S. L. (2004). *Shock and aha! A transformative model to internationalize undergraduate courses.* Workshop presented at the POD Network Annual Conference, Montreal, Quebec, Canada.

O'Donovan, K., & Mikelonis, V. M. (2005). Internationalizing on-campus curriculum: A faculty development program to integrate global perspectives into undergraduate course syllabi. In L. C. Anderson (Ed.), *Internationalizing undergraduate education: Integrating study abroad into the curriculum* (pp. 91–95). Minneapolis, MN: Learning Abroad Center.

Reese, M. (1985, May 10). A tragedy in Santa Monica. *Newsweek,* (10).

Smith, K. A. (1993). Designing a first year engineering course. In M. E. Schlesinger & D. E. Mikkola (Eds.), *Design education in metallurgical and materials engineering: Engineering design in courses and curricula* (pp. 59–73). Warrendale, PA: The Minerals, Metals & Materials Society.

Smith, S. L. (1998). Ethics in intercultural and diversity training: A guiding perspective. *Journal of the Northwest Communication Association, 26*(2), 25–39.

Weisman, E. M., & Garza, S. A. (2002). Pre-service teacher attitudes toward diversity: Can one class make a difference? *Equity & Excellence in Education, 35*(1), 28–34.

Wiggins, G., & McTighe, J. (1998). *Understanding by design.* Alexandria, VA: Association for Supervision and Curriculum Development.

CHAPTER 6

Teaching Technical Communication to American Students in a Study-Abroad Program

Deborah C. Andrews and Brent Henze*

Studying abroad has become an increasingly attractive strategy for American students as they develop intercultural competencies to achieve personal and professional success in the global economy. The Institute of International Education reports a 144% increase in American student enrollment in study abroad from 84,403 participants during academic year 1994–1995 to 205,983 participants 10 years later (Institute of International Education, n.d.). Once limited mainly to students of foreign languages, such periods of study abroad now attract students across a wide spectrum of majors, including students of technical communication and, more broadly, professional writing.

That approach makes sense. Friedman (2005) provides a useful and often-cited analysis of the transformed environment for business, and thus for professional communication, in the 21st century. He sees the transformation as occurring in three phases. In the first, Globalization 1.0, *countries* sought out a global economic stage following Columbus's opening of trade between the old and the new worlds. Religious or imperialistic (or both) motivations caused exploration; natural and human energies powered it. From 1800 to 2000, a period Friedman calls Globalization 2.0, *multinational companies* were the key drivers of a global

*Authors' note: portions of this chapter were originally published in *Business Communication Quarterly,* 72:1 (March 2009). We are grateful to the Association for Business Communication for permission to reprint that text.

economy. Transportation technology (railroads, steam engines), communications technology (telephone, telegraph), and a range of financial instruments and institutions led to extensive international economic growth. Around 2000, however, according to Friedman, we entered a new era, Globalization 3.0. In this, not nations or companies, but *individuals* are the main agents of change; ubiquitous, cheap telecommunications have created a level playing field (thus, the earth is flat) and opened business competition to "every color of the rainbow":

> It is now possible for more people than ever to collaborate and compete in real time with more other people on more different kinds of work from more different corners of the planet on more equal footing than at any previous time in the history of the world. (Friedman, 2005, p. 8)

The nature of work has thus been transformed, with fewer hierarchies, reduced supervision, constant change, and the integration of work across disciplines, nations, cultures, and agencies. Individuals need to deal with other individuals worldwide, often through virtual teamwork among geographically distributed workgroups. Everyone, not just those in an international unit of a company or a government's foreign office, must operate in an international setting.

NEW DIRECTIONS IN STUDY ABROAD

To enhance students' abilities to thrive as professionals in the environment of Globalization 3.0, faculty have recognized that the study of foreign cultures in one's home classroom is not enough. Students need to experience the cultural dislocation that comes with fieldwork abroad. To accommodate this need, U.S. colleges and universities are developing programs for delivery abroad that are innovative in both their time frame and their content. This tendency has multiple implications for teachers of professional writing (a term we use in this chapter to include technical, management, and business communication). Promoting intercultural and international communication competence has long been a goal in professional writing courses and programs, but accomplishing that goal in a study-abroad setting both increases opportunities and introduces problems (Starke-Meyerring & Andrews, 2006).

Students in business and technical fields may find that their tight curriculum and work obligations make studying abroad an obstacle in their progress to the degree. In addition, U.S. administrators are adding fees and other financial requirements as they see revenues drop when universities abroad or programs run by other universities draw off students through lower tuition and housing costs. Some universities require students to pay full at-home tuition even if the host school costs less and cap the financial aid available for study abroad (Colleges impose restrictions on study-abroad programs, 2006). To meet the needs of both students and administrators, international program offices at many schools encourage their own faculty to develop short courses in a variety of subjects, including technical and business ones, that can be taught abroad. In fact, according to the Institute for International Education Open Doors 2006 report, the majority of students (56%) chose to participate in courses lasting less than one semester, the largest growth area; 38% studied for a semester, and only 6% for a full year.

These shorter courses offer many advantages for students as well as faculty. Students who would otherwise not be able to afford the direct costs (program fees, food, and lodging) and indirect costs (lost wages from work) of a full-year or full-semester program may be able to participate in a shorter program; so can faculty who might otherwise not be able to take the time away from their regular semester teaching responsibilities. Instead of sending their students abroad to study at other institutions, these programs provide ways for faculty at U.S. institutions to accompany their students abroad, enabling greater university control over both revenues and curriculum. Developing courses in this context is a complex endeavor that requires readjusting goals and assignments and carefully balancing time in class with time for students to independently pursue the differences that should strike them as they live in a new culture. Career-oriented, study-abroad short courses, however, are rarely discussed in the literature, aside from a few news articles (e.g., Rooney, 2002; Study abroad: The short course, 2004). This chapter attempts to address that gap in our understanding. In doing so, we take as given the likelihood that longer periods of immersion in a culture are better than short ones, that the learning goals for a short course must be more limited than those that can be achieved in a longer course, that simply being in a new culture absorbs unproductive orientation time that may be better amortized over a longer period, and similarly, that settling into a culture helps overcome the "gee whiz" and "on vacation" factors that may continue through most of a short period.

With those cautions in mind, we describe four study-abroad courses developed by professional communication faculty. Our approach is largely descriptive, although we present recommendations based on these descriptions at the end of the chapter. To teach these, the faculty member had to recruit an adequate number of students at the home institution to meet that institution's criteria for financial viability and assume responsibility for designing and implementing the course.

BUSINESS COMMUNICATION IN ITALY

Recruitment was a major concern of a University of Delaware linguistics professor who several years ago centered a short program for Italy around his teaching of the university's basic business communication course there. He knows Italy well, speaks Italian fluently, and enjoys visiting there frequently. The course was an excellent marketing device because it is difficult to enroll in on campus (students were guaranteed enrollment if accepted into the Italy program); the study-abroad ambiance promised students a more relaxed approach; and grades in study-abroad courses are at least rumored among students to trend higher than for on-campus versions. The Italy program is one of the most popular offered by the university's long-term and highly regarded international programs office, regularly enrolling around 60 students, and a significant revenue producer.

For winter term 2006, Andrews signed on to teach the business communication course; the linguistics professor taught a linguistics course that provided the other three credits needed for the program. Andrews reviewed his syllabus, which seemed to make few accommodations to the environment abroad and was considerably less demanding than the course on campus. To some degree, this chapter grew from Andrews's frustration in trying to make the course more rigorous than in its previous iterations and more tied to the local scene and local texts.

As designed by the linguistics professor, this program was a moving one: 8 nights in Rome, 4 nights in Siena, 3 nights in Venice, 7 nights in Florence, 4 nights in Naples, 3 nights back in Rome before flying home. Two buses (one faculty member for each) were used for travel between cities and for tours in each city. Priority during the day was generally given to the tours (museums, churches, historic sites, and the like); the students and Andrews had only 2 days during the period when they were not either traveling or touring as a group (although some tours were only a few hours long).

Given this perpetual movement, it was difficult to fit in the major elements of the course: the employment search; various business genres (e-mails, memos, letters, report); writing in teams; and some kind of a research-based report. To accommodate the first of these features, Andrews held a session with all students on campus at the end of the fall semester. Students then e-mailed their letters and résumés to her in advance of the trip, and she returned them electronically, matching the same standards as would apply to the course on campus.

For the other key elements, Andrews framed assignments specifically for on-the-ground evidence gathering in Italy, especially in a major team project. She took advantage of their role as tourists in what may be one of the most significant businesses of Globalization 3.0, cultural tourism, to prepare students to look critically at the communication aspects of the sites they visited: museums, historic homes and monuments, gardens, and other locations that display objects and interpret them for various publics. The brochures, wall labels, acoustic guides, and scripts presented by living guides were all considered worthy of study as genres of professional communication. In addition, she encouraged students to look behind this display and interpretation to the organizations supporting them. These organizations have goals and visions for themselves; market themselves to a variety of audiences and derive and compete for funding from a variety of sources; exist in an economic, political, and social context; and communicate internally and externally to remain viable. These institutions and communication products thus provided students with a rich resource for study. Andrews also alerted students to the possibility of research projects focused on the restaurants and hotels they encountered. Because several students were majors in fashion and apparel studies and others who were not majors were still interested in the fashion industry and in design in general, retail establishments and markets became sites for field work and cultural comparison. One further advantage of studying such sites was the common use of English in their communication, so those students who did not speak Italian—the majority—were still able to interview locals and read texts. Students used a short, internationally oriented text that fostered discussion and writing exercises and were asked to set up blogs to share their observations, both in text and in visuals, with classmates as well as friends and family at home.

Classes were usually scheduled in the afternoons at times when hotel facilities (like breakfast rooms and lobbies) were available and on 3 of the 4 Saturdays during the period. In one city, classes met for 2 days in the local university, but that was unusual. Classes were packed intensively into 4 of the 5 weeks, a total of 15 classes of 1 hour and 15 minutes each. They met on a revolving schedule with students alternating between the linguistics and professional writing courses.

The students' only access to technology was either in the hotels, which tended to charge business rates for connect time, or at various commercial Internet points. Students who were used to highly sophisticated access, without charge, on campus, were dismayed by this situation. While many commercial sites were clean and well-lighted, many others, especially those clustered near train stations, a common location for such sites, were unkempt and crowded. Students did use them, but highly reluctantly. Limited and somewhat scary access to technology made it difficult for students to work on their blogs and in other ways became a significant detriment to the course.

Class size presented further problems. The excellent physical condition of classrooms, comfortable computer labs, and good study facilities allow a class of 25 to be workable on campus. Absent those conditions in Italy, students were easily distracted in the two large sections of the course, with 24 students in one and 30 in the other. It was hard to give them the individual attention they needed under these odd conditions.

Students also chafed at being in class in the first place. They asked why class was interrupting their experience of seeing Italy, and assignments, which aimed to enhance their understanding of their new environment, became points of contention rather than connection. In the end, many students did improve their understanding of organizational communication processes and products, but the structure of the course did not map well onto their expectations based on rumors about earlier iterations of it in Italy. Most had signed up for the program in order to travel in Italy; they were not primarily interested in business communication and that lack of motivation—an issue on campus as well—loomed large in this environment of even more dazzling options. As a final note, the linguistics professor invited relatives of the students to join the program for the last week or so to fill any open spaces on the buses. A substantial number of parents, grandparents, and siblings accepted his invitation. Their presence provided yet another distraction during the period when students were finishing their projects.

BUSINESS COMMUNICATION IN FLORENCE, ITALY, AND IN OXFORD, UK

A more successful approach to a course similar to the Delaware one is that developed by David Blakesley for Purdue University students in Oxford and Florence.

In recruiting potential students, Blakesley notes in his course syllabus:

> Florence is the birthplace of the Renaissance, which brought forth not only great works of art but also new ideas for business and global commerce. Art becomes a commodity, and ever since the Renaissance, enormous resources have poured into making art into a big and lucrative business. In the Florence context, students will study the rhetorical principles and writing practices that foster effective communication. They will conduct field research in museums and consider how Florence and the Tuscany area are constructed through visual and verbal media as a commercial, tourist culture. Students will also read and discuss Shakespeare's *The Merchant of Venice* and some brief excerpts from Dante's *Inferno* to consider the ways that business and art might form ethical alliances.

The 6-week course, taught in May/June 2005, included daily writing exercises at the start of each class, focusing on readings, films, or excursions; a résumé, cover letter, and an assessment document about employment communication; a field-based project; and a client-based project. For the field-based project, completed individually, students assumed the role of ethnographers as they researched and analyzed marketing strategies used by art-related venues like museums. Students took pictures (they were strongly urged to bring digital cameras), interviewed guides, collected printed materials, visited museum Web sites, and then wrote their analysis of verbal and visual elements in accordance with guidelines provided by Blakesley. In effect, they created their own photo essays on how art functions as a business. The language barrier, of course, posed some problems, but many tourist sites offer at least some of their interpretive guides in English. For the client-based project, students worked in teams to prepare a multimedia recommendation report that involved field research and document analysis.

To support its activities in Florence, Purdue contracted with the Centers for Academic Programs Abroad (CAPA) for a facility with classrooms, a computer lab, and a shared office for faculty. CAPA also arranged for students and faculty to stay in apartments scattered throughout the city and hired adjunct faculty to teach courses in, for example, art history and Italian culture. Some 60 Purdue students participated in the program. Each student needed to take a total of six credits from an array of offerings including those provided by CAPA, a psychology course offered by another Purdue instructor, and Blakesley's business writing course. In addition to study space in their apartments and at the CAPA facility, students also took advantage of an arrangement by which they could use the British Library in Florence.

Blakesley taught one section of the course, enrolling 18 students. It met 4 days a week at the same time each morning for 2 hours. Students were free to engage in their own pursuits for 3-day weekends and at other times, except for two arranged day-long tours outside the city and some shorter excursions to visit sites in Florence. The Study Abroad Program itself served as the client for one student project, including a recommendation report regarding the use of an open-source content-management and blogging system.

As Blakesley notes, one goal of the program is to "immerse students in this experience of living abroad, even the hassles of travel." Students made their own arrangements for getting to and from Florence, found their way around the city, shopped and cooked, and in general lived their own lives outside of class.

In a similar time frame in 2004, Blakesley taught the business writing course in Oxford. CAPA again managed the program, but the facilities for class meetings, computers, offices, and housing were arranged with Oriel College at the university, which makes the space available for such use during its summer term.

Because there was no language barrier as in Italy, students were able to find their own clients for projects. Teams developed, for example, marketing materials for a local restaurant and a tool kit for kids to be used by a local museum. Purdue students merged easily into the local community around their class times, which were similar to those in Florence: 4 days a week, 2 hours a day, with weekends free except for 3-daylong tours.

"DOCUMENT DESIGN AND VISUAL DESIGN," LONDON

Unlike the basic business communication courses at Delaware and Purdue, offered in multiple sections to meet the needs of multiple audiences across the curriculum, Brent Henze's "Document Design and Visual Design" was specifically designed for East Carolina University (ECU) English majors. It is one of three courses offered in an ECU program in London. The iteration of the course described here lasted 3 weeks and enrolled 13 undergraduate and graduate students.

Study-Abroad Adaptation

The course, which was conducted, whenever possible, in the field rather than in a classroom, aimed at learning the principles and practices of design from various perspectives: visual design, document design, information design, landscape design, and exhibit design. As their major project, students researched, designed, and produced a visitor's guide focusing on a thematically unified set of London attractions. The assignment leveraged students' own experiences of disorientation. One student, a World Cup soccer fan, wrote a guide for American soccer fans traveling to London who wanted practical guidance about what to see and do in the city as well as the flavor of London as a soccer city. He conducted his research by visiting all the soccer stadiums within a tube's ride of his flat, talked to fans of various teams, visited a team-related pub during a match to get a feel for the fan experience, and conducted additional research about World Cup teams on the Internet to flesh out the document.

Through the project, students performed the dual roles of visitor and communicator, imagining what others in a similar situation would need to know, see, and do; experiencing the city actively, reflectively, and purposively, not just affectively.

In addition to their own individual project work, students as a group integrated assigned readings with excursions. After reading a chapter on principles of spatial relationship in document design, students visited the Royal Botanic Gardens at Kew, where they were challenged to think analogically about the applications of design principles to physical environments and documents. A guided group analysis of a selected garden space introduced an exercise in which teams of students conducted a close reading of a space of their choice in terms of its purposes, potential audiences, and design elements and then reconvened to discuss their findings. Similarly, the reading of an article about the goals and challenges of museum exhibit design preceded a visit to the British Museum, where students conducted a design analysis of an exhibit of their choice based on a rubric Henze provided.

Such excursions gave students a specific, bounded exercise in which to understand design principles *in situ,* encountering the messy complexities that shape users' experiences of designed objects and spaces. Mirroring Kolb's "learning cycle" model, students experienced an environment, then reflected upon that experience, and finally analyzed the experience according to new concepts taught in the course (Montrose, 2002).

Students and faculty were housed for the entire 3 weeks in flats. They became part of the local scene in their multiethnic neighborhood (Bayswater) and in London, shopping at local markets and making conversation with neighbors. After only a few days of anxious forays, even the most cautious were able to navigate London Transport (the Tube) and know which buses to take (and which to avoid).

Classes met in the common room of the student flat rather than in a separately booked classroom. Students met there for 2 to 4 hours on about half of the 15 weekdays but also met for planned full-group discussions and exercises during all of the group excursions. The small number of participants raised program costs for individual students but had many advantages, including the case of traveling together using public transportation and obtaining inexpensive day-of-show theater tickets. faculty could also individualize instruction, which was particularly helpful since students came to the program with widely varying levels of experience with communication courses.

BUSINESS COMMUNICATION IN INDIA

The courses designed and implemented by Andrews, Blakesley, and Henze brought students to traditional sites for study abroad—the UK and Italy. A course designed by Mahasveta Barua of the English department at the University of Delaware brought students to an emerging site of interest in professional communication: India. In doing so, it addressed special challenges in introducing students to a developing country.

India provided an attractive site for such study for a variety of reasons. The professor is from India; India is an emerging force in the world of business, entertainment and literature; and, most important, English is one of the official languages of India and the language used in business, legal, and academic communication. The course was based in New Delhi, but students traveled over a two-week period to the northeast of India—to the state of Assam, and to Darjeeling in the state of West Bengal, and later on a two day trip to Agra to see the Taj Mahal, and the palace city of Jaipur. Though Delhi is not as important a corporate city as Bangalore or Mumbai, it is culturally diverse and rich in history, and it has a large student population with its numerous colleges and universities.

The idea behind travelling to relatively remote parts of India was that if students were to experience India, they needed to experience the diversity that defines India and not be confined to major cities alone. Staying in Delhi for the first week and the last two weeks, however, allowed students to become familiar with the city. By the last week they knew the rates for taxis and scooter-rickshaws, negotiated fares with ease, and even travel on the Delhi Metro system. They quickly discovered which restaurants had the best prices and food and how to get to the various museums and art galleries. The one firm rule was to never go out alone, and students abided by this. Barua had also introduced them to many carefully vetted, local students with whom they became good friends. This served her and the students well; they were able to enjoy movies or clubs in the evenings with their new local friends (with whom they remain friends), and she did not have to chaperone them to these.

In addition to the business communication course, Barua also taught a course on women writers to provide the 6 credits necessary for a full study abroad program. Enrollment was kept to twelve students, and by sheer chance there were equal numbers of male and female students. In the semester before their winter term program Barua warned students to expect the unexpected while traveling in

India and to 'go with the flow' as is the Indian custom. But one major unexpected circumstance arose more than a month before the trip even began: the terrorist attacks on Mumbai in November 2008. Suddenly, the trip itself was in jeopardy and anxieties ran high, especially among parents and University administration. Following the US State Department's travel advisory, Barua provided a detailed itinerary and description of safety measures that would be put in place, and the program was not cancelled. Students were given the option of withdrawing without penalty if they chose; only one student dropped out. Eleven students and one faculty member, an even dozen, proved to be just the right number for ease of instruction, project completion, and daily travel.

Barua developed the courses so that students would not feel a disjunction between writing business reports and doing literary analysis. To that end, the short stories and novels that were part of the women writers course considered how the cultural environment impacted the position of women in various time periods. For the business communications course, Barua planned visits to a number of American and multinational corporations and meetings with both male and female executives so that students could compare the lives of the women they met with those they read about. She also arranged for students to interview business women and some writers to gain information for their business reports (as regards women's position in the workplace, communication styles, dress codes, and the like) as well as their literary responses. She also assumed that they would be making constant observations as they moved through the city and rural areas that would inform their writing in both courses.

For the business communications course Barua chose an online text with many online exercises and short tests; the women writers course included translated texts bought at an organization—an NGO—that focused on education and translation. This choice of texts resulted in completely mixed outcomes: the online text idea proved to be a disaster, whereas the translated literary texts led to an incredibly fruitful connection with the organization that published them.

One of the most difficult elements in this program was access to computer facilities and the Internet. As Andrews notes about Italy, there are numerous commercial internet sites in Delhi and other cities in India, too, the conditions of which are not ideal for prolonged use. The students stayed at the YWCA's International Hostel in Delhi, and though this had limited technological facilities, Barua was assured at an earlier visit to the site that it did have Internet access. But when the group arrived at the YWCA, one section of the building was undergoing massive renovations and extension, and there was no internet facility available for the month. The students were directed to the adjacent YMCA, but they soon discovered that these organizations drew clear lines in the sand and were not very welcoming of each other's guests. The idea of doing online exercises everyday had to be abandoned; however, these constraints formed a point of discussion for the business course. The students quickly realized the need to select communications media dependent on material factors and develop communications strategies and alternatives based on what was available. Luckily Barua had PowerPoint slides of all the chapters of the online text on her laptop, and the business class was conducted sitting around the breakfast table, with the laptop visible to all. These breakfast meetings were very productive; students were able to discuss ideas, ask questions of the teacher and each other in a casual yet very contained setting, and they all later commented on how much they enjoyed these meetings. Students also learned the value of developing close relationships with employees of businesses one patronized, especially within a high-context culture like India. Everyone in the program developed good working relationships with the YWCA staff, and the staff allowed Barua and her students the use of the dining room even after it was closed for meals. The discussion for the literature course took place in various settings. Students read on their own, of course, but very often they took turns reading the assigned poems and short stories out loud while traveling in the bus. Reading out loud, or telling/narrating was the governing idea of the organization—Katha—that Barua and her students visited to purchase the literary texts.

Katha is a twenty year-old NGO that focuses primarily on educating slum children through storytelling, with its secondary focus being translating regional language fiction and non-fiction. On the first visit to the organization's office, students met and talked with the board of directors, and on a later visit they met with the founder and executive director and visited and observed a school in the slums. Students had been formed into three groups for their final project for the business communications course, and one of them chose to work on developing a possible link between Katha and the University of Delaware. That project was very successful, and the student group continued to work on it even after the program was over and through the spring, when they presented it to University administration before putting it into action in the next year. The team also hopes to return to India to implement their proposal. The fact that students were able to take a proposal beyond an assignment and make it into a real-life project is one indicator of the program meeting its goals.

The assignments for these courses varied. For the communications course students produced writing similar to what Barua assigns to her class on campus: an investigative report on the communication needs of their chosen field for which they interviewed someone in that field with their own career preparation in mind; an evaluation of persuasive documents that they collected during their visits to historical sites, museums and art shows; a report on their observations of oral persuasive communication as they reflected on the strategies of the many hawkers and hustlers they met. They wrote memos on their progress on various assignments and produced a resume. As noted earlier, for the final report, teams of students wrote proposals to an aid-granting agency or other organization to support a humanitarian project, or devised a plan to raise funds for such a project. Since these were based, as with the Katha visit, on what they encountered while traveling or going about their daily business, their time was well used. The other two groups wrote final proposals on a sustainable plan to reduce the large population of child beggars on the streets, and a feasible method of dealing with the garbage problem in Delhi – both problems they faced every day of their stay in New Delhi. They were able to survey locals for their opinions and interview some government officials. Part of the trip to the east of India was spent in home stay, and students had access to local authorities through their host families.

The assignments for the literature course consisted of critical responses to the readings and maintaining a journal through posting blogs about their travels, which they did whenever they had access to the internet. Rather than emailing assignments, students simply saved them on their flash drives that they handed in periodically, and Barua was then able to read and grade them. The final group assignments were turned in two days after the group returned to the U.S.

According to Barua, the best part of planning this program was the ability to individualize certain projects and experiences. Dealing with such a small number of students and knowing their majors and interests allowed her to plan the travel to the eastern states carefully. The trips to Assam and Darjeeling allowed students a first-hand look at the tea industry – from plantation, to packaging, to marketing – tourism and travel, and hotel and restaurant management. The visit to the wildlife sanctuary (Assam is the home of the Indian rhino), while being fun for all, was especially useful for the Wildlife Conservation major. For the benefit of the two fashion majors, students visited silk-weaving facilities; however, everyone enjoyed meeting and buying directly from the weavers with no middleman involved. This trip also allowed students to look at the background of Assamese women and meet with Indira Goswami, the author of *The Man from Chinnamasta*, a novel about animal sacrifice set in the ancient temple of the mother goddess Kamakshya. Before meeting the author, the group read the work and visited the temple; a brave few even decided to watch a goat being sacrificed. Each student was asked to have ready at least one question about the novel or related issue during the meeting with the author. This was a tremendously successful exercise, as the author appreciated their in-depth reading of her novel, and students were able to get an unique author's perspective on the text.

The visits to the businesses were helpful but could have been better planned. The best was the afternoon spent at the Indian Oil Corporation (IOC), where the Director of Communications discussed such important issues covered in the course as internal and external communications as well as understanding media and targeting an audience. IOC is a Fortune 500 corporation that functions autonomously; however, it is a Government of India entity with its CEO holding a post similar to a cabinet secretary's. The Communications Director's talk enlightened students about what a corporation reveals to and conceals from the public, and how he himself maintains strong communications ties with the public, the press, and the shareholders. Since the oil industry had been on strike the week that Barua and her students arrived in Delhi, a strike that nearly brought the country to a standstill, students were able to ask really pertinent questions about how his department worked to put a positive spin on situations like that. On a visit to the Indian headquarters of Macy's, the department store, the general manager allowed students to see how communication takes place across the globe and how one idea, originating in New York, has to be interpreted consistently by groups in five different countries in Asia, all speaking different languages. The one problem with these visits was scheduling. There were too many official holidays during the month of January that Barua had not factored in, and these site visits were not always consistent with what students were working on, though they were originally meant to be. However, this was more an inconvenience and not detrimental in any way to the learning process.

Overall, this first run to India can be said to have been successful, given all the things that could have gone wrong. India is a country where plans must at best be tentative. Power outages, traffic, fatigue due to weather and crowds, a lax attitude towards schedules, sudden closures of roads and businesses due to political unrest, all play havoc with plans. But accommodating these factors was an important element in the students' learning. The trip from Assam to Darjeeling was postponed by a day and was on the verge of being cancelled because of a strike and *bandh* (or closure) in that district. The tour agent who had booked the domestic tickets, however, immediately contacted a local guide who traveled down to test the safe routes and get assurances from local authorities; he contacted the agent, who offered to either change all the travel and hotel bookings by a day, or arrange a trip to an alternate Himalayan town at no extra cost to the program. Barua was able to use this as an example of how competitive businesses use technology, people on the ground, and quick planning and viable alternatives to keep customers and foster goodwill. And students noticed other elements, both good and bad, that Barua had not pointed out – inordinate care and hospitality from all the organizations they patronized; instant connections and friendships with people who served them, from the laundryman who brought their clothes to their door to the taxi driver who waited hours while they shopped; a consistent attempt by merchants to inflate prices when dealing with westerners; aggressive selling by souvenir shop vendors at tourist spots in the north as opposed to the laid-back attitude of vendors in the east, with students learning how and when to say no; the unbelievably low cost of skilled labor that they availed themselves of, from having clothes custom made to getting shoes mended. By the last day all the students felt that though they had not stayed long enough, they had stayed long enough to be able to go back alone to India and communicate with ease.

RECOMMENDATIONS FOR STUDY-ABROAD COURSES IN PROFESSIONAL WRITING

As many researchers have noted, the field of professional communication has increasingly recognized the cultural specificity of its practices and thus the need to educate practitioners to thrive in an environment of those differences. One attractive strategy for meeting that goal is taking students abroad, immersing them as time allows in the new context, and using that context to enhance their learning. The short time frame is clearly a limitation to these courses, as is the fact that students from one university often live together in a kind of island of that university abroad. Even given such circumstances, however, the courses can succeed in enhancing students' learning about the situations and genres of professional writing within a framework of enjoying a new culture. In this section, we provide a few recommendations for doing so.

Settle in One or a Few Locations

Too much time on buses or trains, too much packing and unpacking of suitcases, and too much need to orient to new streets, new restaurants, new Internet points are activities that divert energy from more meaningful, higher-order thinking and writing. Students need time to be both excited and bored and a regularity of schedule that fosters study as well as recreation. While some travel and home-stays can be enlightening, especially in developing countries as in the India program, a road-trip approach only reinforces the notion that students are on vacation and not in a learning environment. Students gain from living on their own and engaging with their neighborhoods, a practice that proved highly successful in the Purdue and ECU programs.

Advocate for Small Class Sizes and Recognize the Business You're In

The argument for small writing classes is an ongoing one in the United States and seems particularly pertinent to professional writing taught as a short course abroad. Andrews attributes at least some of the problems in her classes to their large size; Blakesley, Henze, and Barua seemed to reap benefits from their much smaller groups.

In designing a study-abroad version of a professional writing course, however, faculty need to recognize that they, too, are in a different academic country. Most such programs are offered through special offices (international programs, special sessions, continuing education) that are more entrepreneurial than traditional academic units. Thus study-abroad programs must meet profit-and-loss criteria as well as net-income goals. Faculty members should not balk at these conditions but instead respond to them creatively to achieve academic objectives. For example, to keep class size at educationally appropriate levels, faculty may suggest reducing program costs (like admission and bus costs that were bundled into the University of Delaware program fee) to maintain the institution's net margin. A program with fewer students can still yield a profit if costs for nonclassroom activities are not paid from fees. Faculty need to be flexible and entrepreneurial themselves and work cooperatively with the sponsoring office to meet both financial and educational objectives.

Foster a Resident's Rather Than Tourist's Perspective

The road trip model of the Delaware program in Italy tended in some respects to infantilize students, who at the same time expected not to have to apply individual initiative and rebelled against the rigid and seemingly random requirements of tours and classes. In addition, the 24-hour, 7-days-a-week access that students and faculty had to each other in that program (they stayed in the same hotels, usually on the same floor) served some students well but also created a bubble that kept the new culture at bay for many of them, except at the most superficial level. Even in longer-term programs in which students travel individually to study abroad, there is a danger that students will gravitate to the relative safety of other American students or the nearest expatriate pub rather than risking themselves by attempting to experience the local culture more fully. In short programs, it is particularly important for faculty to encourage (or compel) students to explore the place and interact with the culture on their own, even though they have less time to learn the ropes (Langley & Breese, 2005). Student shoppers need to focus less on items to be brought home and more on those they consume as a natural element in their lives abroad.

Explicitly Accommodate Technology

Excellent access to appropriate technology underpins professional writing courses. An important step in planning such courses for delivery abroad is thus planning course technology. Contracting with on-site organizations like CAPA or with local universities may meet this need, although any arrangements with such providers must be transparent, the partnering organizations tested for reliability and ethical behavior, and the costs appropriate. Blakesley also arranges a collaborative Webspace through the Purdue server. Henze phased his course so that students researched their major project during the three weeks abroad but then completed the documentation and production during a final 2-week phase on campus where they had appropriate access to technology. Internet cafés are common and relatively cheap in large European (and other) cities, but they should be considered supplemental rather than essential sources of access.

Blend Formal and Collateral Learning

Although students need the discipline of regular class meetings and regular assignments, such meetings and assignments only enhance their feeling of being students who happen to be abroad. Class time provides structure and targeted conversation, but collateral learning may be the main point of the program. Such learning can be maximized with judicious but modest use of exercises, framing discussion, and other opportunities to reflect.

Another approach to minimizing class time abroad is to hold classes on campus before the international stay, as is common at Delaware. A Delaware photography course conducted in Australia over the winter session, for example, met on a sequence of several Saturdays during the preceding fall semester; a finance class met nine times during the fall as students developed background papers and presentations on companies whose offices they would visit in Geneva in January. This approach, however, may encounter understandable resistance from students already carrying a full load of courses during the previous semester. At a minimum, students may be required to read texts and background articles in advance.

As we've seen with each of these courses, faculty felt it important to use a problem-based learning approach to instruction, which redefines what it means to be "in class" in general (Brandt & Manley 2002; Katula & Threnhauser, 1999; Steeves, 2006). Project time counts, including team meetings and instructor-team consultations. Agencies like CAPA, international organizations like the Society for Technical Communication, as well as school and personal connections can help faculty find potential project clients abroad as well as interview subjects in local as well as multinational companies. Client projects are most appropriate when the course is set in an English-speaking country. It's much more difficult to do so where students are not fluent in the local language. But it's possible in such a setting, as Blakesley has shown, if students stay in one location so that they can adjust to its particularities and have time to develop local contacts.

Grow Local Roots

Obviously, the course needs to grow local roots. Locally based and client-based projects are one good approach. Helping students see museums, shops, and hotels as sites of professional communication can be enlightening, as can the dislocation of studying abroad, which then becomes part of the course content. In his course, Henze challenged students to see the centrality of culture in all design practices, both productive and analytical, and thus to consider their cultural assumptions about design and to imagine what different cultural assumptions others might apply. The framework of design enabled students to examine the culture-specific objects, expressions, and places that they encountered; and the cultural differences that students discovered in this environment, conversely, helped to denaturalize or make strange aspects of design that students might take for granted when they encounter design in a more familiar environment.

Strike the Right Balance between Holding On and Letting Go

Faculty who develop a professional writing course to be taught to their American students abroad have to strike a fine balance among competing elements: classes, assignments, tours, and free time. They have to both structure the experience and let the students learn as they go. Obviously, a home course cannot simply be packed in a suitcase and delivered whole to that setting. As we hope to have shown in this chapter, developing short courses in professional writing can be complex. But the process can also be engaging and rewarding, both for faculty and for students.

REFERENCES

Bernstein, E. (2006, April 12). Colleges impose restrictions on study-abroad programs. *The Wall Street Journal*, p. D1.

Brandt, C., & Manley, T. (2002). The practice of the fieldbook: Facilitating and evaluating field-based learning. *Frontiers, 8,* 113–142.

Friedman, T. (2005). *The world is flat: A brief history of the 21 Century.* New York: Farrar, Straus, and Giroux.

Institute of International Education. (n.d.). *Open doors 2006: U.S. students studying abroad.* Retrieved April, 11, 2007, from http://opendoors.iienetwork.org/page/89252/

Katula, R. A., & Threnhauser, E. (1999) Experiential education in the undergraduate curriculum. *Communication Education, 48,* 238–255.

Langley, C. S., & Breese, J. R. (2005). Interacting sojourners: A study of students studying abroad. *Social Science Journal, 42,* 313–321.

Marklein, M. B. (2004, February 2). Study abroad: The short course. *USA Today*, p. D1.

Montrose, L. (2002). International study and experiential learning: The academic context. *Frontiers, 8,* 1–15.

Rooney, M. (2002, November 22). Keeping the study in study abroad. *Chronicle of Higher Education,* A63.

Starke-Meyerring, D., & Andrews, D. (2006). Building a shared virtual learning culture: An international classroom partnership. *Business Communication Quarterly 69,* 25–49.

Steeves, H. L. (2006). Experiencing international communication: An internship program in Ghana, West Africa. *Journalism and Mass Communication Educator 60,* 360–375.

CHAPTER 7

Global Revisions: (Re)Thinking the Future of Technical and Professional Communication Competencies

James Melton

In the literature on international professional communication, it has almost become an opening ritual to note that the global context presents new challenges for professional communication programs and practitioners. Ideally, the field would already have identified how educational practices and curriculum should be adapted in response to these changing conditions. While strides in this direction have been made, the gap in our knowledge has persisted, partly because conditions are changing so quickly. Perhaps an even bigger obstacle is that we have yet to take up Timothy Weiss's (1997) charge to examine our basic assumptions about the role of the professional communicator and the principles of professional communication in international context. This chapter describes five areas of competence for international professional communicators drawn from a case study I did with an international training team in Japan (Melton, 2008b). I explore how each area is revised when the common U.S.-American assumption of information sharing is replaced by the common Japanese assumption of relationship building as the primary purpose of professional communication.

ASSUMED ROLES AND PRINCIPLES OF PROFESSIONAL COMMUNICATION

Weiss concludes his 1997 article, "Reading Culture: Professional Communication as Translation," by arguing that defining global competencies means redefining the basic assumptions of the field itself:

> To internationalize business and technical communication in U.S. colleges and universities is not simply a matter of refocusing the subject matter; it demands addressing basic yet also advanced skills that are currently hidden by our very definition of the professional communicator and the principles of professional communication. (p. 336)

In the years since Weiss's article was published, the issue he raises still remains: What skills are hidden by our definitions of the professional communicator and the principles of professional communication? To answer this question, we must simultaneously try to identify the necessary competencies and examine our own assumptions. This chapter describes a recent case study of international trainers, in which I explored the abilities international professional communicators actually used. The objective of this chapter is to present the preliminary results of this case study.

Audience Awareness

A common recommendation in scholarly literature and textbooks is that professional communicators should consider audience even more carefully for an international context than they would for a domestic context (Anderson, 2002; Hoft, 1995; Johnson-Sheehan, 2005; Locker, 2003). Certainly, this is sound advice, and the scholarly literature and professional communication textbooks have made great strides in addressing the issue of international audiences. Yet, the advice to simply try harder and gather more information does not get to the core of the issue. We may learn a great deal about international audiences, but without an accompanying look at our own assumptions, we will be hindered in our efforts.

The traditional approach emphasizes data gathering and analysis of one's audience to facilitate good rhetorical choices, but because in international professional communication, we are often talking about radical differences of language, culture, and communication style, to truly be effective, we must learn just as much or more about ourselves as about our audiences. We cannot assume a shared frame of reference; our language, our versions of history, our views of time, and our assumptions about professional relationships may all be different. Of course, there will always be individual differences in these areas, but research in the fields of anthropology (Hall 1989), sociology (Hofstede 1996), human resource management (Hampden-Turner & Trompenaars 2000), and psychology (Schwartz & Sagiv, 1995) shows that there are real and definable differences between cultural groups, while acknowledging that there will be individual variation as well.

A real problem is that this research, misapplied, can lead to stereotyping, a move which must always be guarded against. Yet, an equally negative but more subtle problem is that if, in attempting to defend individuality, we deny the possibility of real differences between groups, we unwittingly make the world into a mirror of ourselves. Focusing exclusively on individual variation allows us to think we are acknowledging difference, while our unstated and often unconscious cultural assumptions continue to remain hidden. This tendency is described by intercultural scholar Milton J. Bennett (1998):

> Many teachers and trainers of intercultural communication find that while most people acknowledge superficial behavioral differences in dress, custom, language, and so on, it takes but a scratch of this surface to encounter a basic belief in the essential similarity of all people. The statement indicative of this belief is, "Once you get used to their different (dress, manners, style), they're just like us!" Attempts to point out more fundamental value differences may even be met by hostility—an indication of how central the assumption of similarity is to our worldview. (p. 192)

It is important to note that while this statement could also be made in arguing against the assumption that all individuals are the same, Bennett is talking here about groups. He explains,

> Unless we can accept that other groups of people are truly different—that is, they are operating *successfully* according to different values and principles of reality—then we cannot exhibit the sensitivity nor accord the respect to those differences that will make intercultural communication and understanding possible. (1998, p. 196, author's italics)

Thinking that we have moved beyond the stage of assuming similarity is perhaps evidence that we are still in it (Bennett 1998). In fact, the tendency to assume similarity is probably not a stage at all, but rather a constant challenge that we must find strategies to deal with, especially when trying to communicate with international audiences. The solution is not to deny differences between groups or individuals, but to examine our own assumptions and acknowledge real differences with respect to both individuals and cultural groups.

On the level of culture, many implicit assumptions about universal ideals for professional communication might be examined, which are, in fact, U.S.-American ideals for professional communication. Some of these common assumptions include the perceived worldwide predominance of American English (Maylath & Thrush, 2000; Melton, 2008a; Weiss, 1999), the valorization of writing over orality as noted by Thatcher (1999, 2000), and the valorization of writing and orality over the unspoken (Yamada, 1997). However, in this chapter, I want to focus on another core assumption that is particularly relevant in many cultures in Asia, including Japan, where I conducted a case study with an international training team.

Information Sharing and Relationship Building

Yamada (1997) shows that the purpose of communication in Japanese professional contexts may be much different from the information-sharing purpose of communication typically assumed in North American professional communication textbooks and literature. Building and maintaining relationships tends to be the primary purpose of Japanese professional communication, whether formal or informal, rather than simply being a nice side benefit (Hampden-Turner & Trompenaars, 2000; Hofstede, 1996; Ramsey, 1998; Yamada, 1997). In contrast, while North American professional communicators might agree that building relationships is an important aspect of communication, they do not tend to give it precedence over the exchange of information (Yamada, 1997); rather, relationships tend to be subordinate to the information that is being exchanged (Hall, 1976; Yamada, 1997).

It is important to note that information sharing and relationship building are not mutually exclusive; in fact, each one facilitates the other. Good relationships often facilitate the efficient exchange of information, and vice versa (Hampden-Turner & Trompenaars, 2000). The question, however, is one of priority, and it makes all the difference. A typical example is that American businesspeople have frequently expected to give a presentation and sign a contract shortly after arriving in Japan. They assume that if they do a good job of making their case and the deal is mutually beneficial, the decision can be made rather quickly. If socializing is involved, they assume it will be an occasion for everyone to celebrate together afterward. To the Japanese businesspeople, this is backwards. They want a few hours or days to observe and interact with their new associates in informal settings. It is here where they can decide if these are people they really want to do business with. Only then will they be interested in hearing what their guests have to say and the deal can be made and the contract signed (Hampden-Turner & Trompenaars, 2000; Yamada, 1997).

In each case, relationships and information are involved, but with different priority. For the U.S.-American businesspeople, if the information is good, then a relationship can be formed. For the Japanese businesspeople, if the relationship is good, then the information can be shared. While this is just one example, the implications of these values can be explored for all aspects of professional communication (Hampden-Turner & Trompenaars, 2000; Yamada, 1997). The case study I conducted showed that relationship building extended to several areas of competency, including language fluency, translation, intercultural communication, and collaboration. The following sections describe the context of international training and the research methods, analysis, and findings of this case study.

Global Training Contexts

Training is a particularly important communication challenge for technical communication scholars and practitioners to consider and an ideal forum in which to observe the use of rhetorical competencies in context. Training can be a vital part of the value-added process, noted by Faber and Johnson-Eilola (2002), who argue that the most desirable future for technical communicators is not merely to turn out information products but to "find ways to leverage our knowledge and build new knowledge to create and add value in a business culture that is increasingly agnostic to physical products" (p. 141). They give an example: "Our task is not to make clear why a group of technical communicators can build a better Web site. Our task is to add value to that software by teaching people and companies how to use it to solve their problems" (p. 141). Technical communicators who have inside tacit knowledge as well as technical expertise can make unique contributions. Their role will be "not merely to learn and teach new technologies, but also to provide creative and innovative solutions to unique and specialized business problems" (p. 143).

A major avenue for approaching these "specialized business problems" is specialized business training, which often complements the organizational role of technical communicators. The practical benefits of combining technical communication and training have already been noted by Hackos (1994), who states, "As a strategic-planning consultant, I strongly recommend, in fact, that training and publications be closely aligned in the development of documents and job aids" (p. 124). She notes that key to both functions is an accurate analysis of audience, which is, in many cases, the same audience. Carliner (2001) adds that technical communication is evolving toward a "campaign-like approach more typical of marketing than of traditional documentation libraries. We can no longer tell users something once—we must use a variety of media and methods to reach them" (p. 156). Customized training can be one of these methods, delivered in a traditional face-to-face setting or through online media, interactive video, print documents, or increasingly, using a blended combination of these (Armstrong, 2002; Bridges, Baily, Hiatt, Timmerman, & Gibson, 2002). As markets continue to globalize, products in any of these media will have to effectively cross national, cultural, linguistic, and technical boundaries or be severely limited in their application and marketability. For this reason, internationalization consultant Marcia Sweezey recommends that training be included as one of four parts of a coordinated information strategy to solve localization and translation problems common in international technical communication (Hoft, 1995). These four parts include printed information products, user interface, online information products, and training.

Because of this increasing integration, technical communicators need to consider how their abilities might fill the need for specialized training in a global context. This is perhaps best learned by finding out what abilities international trainers actually use in their work. The literature has shown that, as with all kinds of technical communication, trainers need to make content accessible to their audiences, which may require not only translation but comprehensive cultural adaptation as well (Huang, 1996; Marquardt, 1995; Thornhill, 1993).

Cultural expectations about methods, objectives, and content of training will need to be considered, as well as assumptions about age, gender, and ethnicity. Choice of training media and methods must be considered in light of linguistic and cultural factors (Connor, 1996; Thatcher, 1999, 2000). It is also highly possible that linguistic or cultural backgrounds will differ not only between trainers and audiences but between training team members themselves. A training team might consist of a curriculum developer in the United States, a multimedia specialist in India, and a face-to-face presenter from Singapore, with an audience in Shanghai. Increased diversity among training-team members may enhance potential, reach across cultures, and add to overall effectiveness, but it may also pose intercultural communication challenges within the team (Hampden-Turner & Trompenaars, 2004; Iles & Hayers, 1997; Lazear, 1998; McDermott, Waite, & Brawley1999).

While this literature has traced the outlines of important abilities that are needed, some questions are still unanswered. Among these are the following: Since most training will be done by teams rather than by individuals, how do international training-team members apply individual competencies in collaboration with each other? How do team members construct and share roles in the use of these competencies? How does the predominant rhetorical purpose of relationship building versus information sharing in some cultural contexts affect the team members' use of competencies? How can the answers to these questions be acted upon by technical communication students and instructors? The research presented in this chapter is a first step toward answering these questions.

METHODS

During the summer of 2004, I conducted research on an international project team based in Honolulu, Hawaii, that regularly delivers sales and management training seminars in Japan for a health food company that is based in the United States. I wanted to find out what competencies the members of the team used, how these competencies related to each other, and how the participants used these competencies in collaboration with each other. For each of these questions, I became particularly interested in finding out how an emphasis on relationship building over information sharing shaped the team members' uses and applications of the competency.

Case Study

Qualitative research, specifically a case study, was the best approach for answering these questions. As Flick (2002) explains, the object of qualitative research is "less to test the already well-known (for example theories already formulated in advance) than to discover the new and to develop empirically grounded theories" (p. 5). The qualitative method of case study allowed me to focus on competencies, which are complex and rooted in particular situations, while still using existing theory as a guide. Sullivan and Spilka (1992) argue that "a case study works from pointed questions as a starting point. This makes the case study more pointed in its inquiry than the ethnographic study" (p. 604).

Pointed questions about competencies were the logical starting point for this study: While studies about intercultural competencies are lacking in the field of professional communication, relevant knowledge has been generated in other fields. This body of knowledge, although not adequate by itself to resolve the research problem, was used to develop questions about intercultural professional communication competence. These questions were more specific than would have been possible if no previous studies had been done (a situation that would have favored ethnography). While survey research might provide a broad view of the competencies individuals believe they and others are using, case study allows for in-depth, observation-based examination of these competencies while maintaining the context within which they are used (Flick, 2002; Yin, 1994). Case study is also well-suited to exploratory research because its data gathering and analysis is an iterative process that takes place both during and after field-work (Flick, 2002; Miles & Huberman, 1994; Yin, 1994).

Participants

The case study involved three participants. The training team leader, Branford (pseudonyms are used for all participants), was a Chinese-American man in his late 50s who spoke English as a first language and Mandarin Chinese as a second language. The second participant was a Japanese man named Okito in his late 30s who was not only bilingual but also bicultural, having grown up in both Japan and the United States. The third participant, Emi, was a Japanese woman in her mid to late 30s who moved to the United States as an adult, where she operated her own English-Japanese translation company for several years. Branford and Okito had been developing and delivering training seminars for Japanese audiences together for about 5 years, and Emi had been a part of the team for 4 years.

Data Gathering

Before the study began, I created a theoretical framework based on my review of the literature on intercultural competence, also described in Melton (2008b), which acted as a guide for data gathering and analysis and was gradually revised throughout the process to reflect the findings described in the results section. This framework, depicted in Figure 1, outlined what I thought I would see, based on the best available theory drawn from several academic disciplines, including English as a second language, training, and intercultural relations.

Using the team's 3-month training cycle as the boundary of the case, I observed the development of a sales training seminar over a period of 11 weeks, from the

Linguistic competence

- Correctly using a recognized, agreed-upon version of language at the appropriate times
- Understanding and interpreting what others say
- Working with translators and interpreters

Intercultural Communication competence

- Recognizing high- and low-context communication and switching between them appropriately
- Choosing appropriate media for communication
- Recognizing and adapting to cultural patterns when interpreting and creating texts
- Seeking and using competent advisors in decisions about discourse in intercultural communication

Nonverbal (including visual) communication competence

- Recognizing and interpreting nonverbal communication by others
- Communicating nonverbally with others
- Seeking and using advice on nonverbal communication from cultural insiders
- Understanding the role of universal elements of human behaviors such as eye contact, turn taking, gesturing, and politeness norms

Social competence

- Self-awareness and cultural awareness: Becoming aware of one's own and others' cultural values and practices
- Sensitivity: Recognizing and reconciling differences in cultural values
- Tolerance toward differences in cultural values and communication strategies
- Interest in and ability to learn about local culture and customs
- Respect: Having and showing regard for individuals and their cultures
- Flexibility: Emotional facility in coping with changes in environment

Adaptive competence (an element in all five of the other competencies)

- Recognizing what skills and knowledge are needed
- Discovering ways to obtain necessary skills and knowledge
- Choosing when to "do it yourself" and when to collaborate
- Assessing the success of your communication strategy and adjusting accordingly

Figure 1. Preliminary model for U.S.-Japan intercultural professional communication competencies.

planning phase in Honolulu to the delivery of the training in Japan (4 full days). The participants understood training as a holistic process that, in addition to communicating information, was also aimed at building and maintaining relationships, which they recognized as particularly important in Japan (Yamada, 1997). Thus, it was not enough to look at the training seminar only; all of the participants' efforts related to the training also had to be considered, including gathering feedback about what training topics were needed, planning social events and recognition awards, and meeting with individuals to answer questions.

My primary means of data gathering were observing planning meetings, interviewing participants, collecting documents, including PowerPoint presentation slides, and observing training seminars and social events. Planning meetings took place primarily at the beginning of data gathering; training seminars and related events toward the end. Throughout the process, I also conducted a total of 10 formal interviews, many of them in a group format with my participants, the three training-team members. My main purpose in these interviews was to check what I thought I was observing in the planning meetings and learn more about it. I also observed the delivery of training in Hawaii, where English and Japanese were used side-by-side throughout, and in Japan. The seminar in Japan was conducted only in Japanese except for Branford's portion, which Okito and Emi interpreted from English to Japanese.

Having only a rudimentary knowledge of Japanese, I was in a linguistic position similar to that of Branford. I was able to understand what was going on in the seminar in Japan because I had already seen the hybrid version in Hawaii and because the team had been reviewing the content with each other in English for several months beforehand.

FINDINGS AND ANALYSIS

Using the preliminary framework discussed in the methods section, I did a rhetorical analysis of each interview, meeting, training seminar, and social event, as well as the overall process of creating and delivering the training. My analysis focused on identifying patterns that would show what competencies the members of the team used, how these competencies related to each other, and how the participants used these competencies in collaboration with each other. Patterns fell into four main areas around which I organized my findings: linguistic competence, translation competence, intercultural communication competence, and collaborative competence.

Language

Linguistic competence is defined as "the ability to produce meaningful utterances" with respect to a given language through speaking or writing (Byram, 1997, p. 10). The case study revealed two important findings for technical communicators entering the context of international training:

1. Using the host language was not just a way to transfer information but a rhetorical act with the potential to connect the identities of the audience and the presenter.
2. The participants' ability to use the host language rhetorically was both constrained and enabled by the cultural expectations of the audience.

Participants' use of the language as a rhetorical strategy to build and strengthen relationships ran throughout the case study. One example was Branford's use of Japanese writing to build rapport with the audience. Japanese written characters, or *kanji,* are borrowed from Chinese and have similar meanings as Chinese characters. Thus, as a Mandarin speaker, Branford was able to write *kanji* impromptu on a white board, usually as Okito or Emi were interpreting his spoken English into Japanese. This act did not serve a practical function beyond the few characters that were written; instead, it served a larger symbolic function by creating a cultural connection between Branford and the audience, a phenomenon also noted by Campbell (1998). Branford explained, "When I *write* in Chinese [they say] this guy is smart or this guy relates to me . . . [so] although I don't do the spoken language well . . . I can do the written and . . . it just closes the cultural gap." All three participants said this ability was a key reason Branford was able to build rapport quickly, which they believed made the audience much more willing to accept and value what he said in his training presentation. A similar process is described by Richmond (1995) as also occurring in Eastern European contexts.

Significantly, participants were aware that cultural expectations influenced how they could use the language rhetorically. With regard to the above example, the participants noted that Branford's use of spoken Japanese would not be perceived in the same positive way. His lack of spoken Japanese fluency, combined with cultural expectations about his appearance, would make such a strategy counterproductive. The participants explained that although Branford is Chinese-American, many audience members in Japan think he looks Japanese or at least familiar, often commenting, "He looks like my uncle." This perception puts limits on Branford's use of spoken Japanese. Okito explained, "If you're a foreigner who speaks Japanese, that'd be awesome, but in Branford's case . . . if he spoke Japanese people would feel like, 'Oh he talks real bad.'" He added, "But because he's talking in English, people perceive, 'Oh that's right, he's American.'" The research participants believed that Branford had a certain ethos as a Western foreigner (Yamada, 1997) and that for this reason, the audience was more attentive to his ideas.

From this perspective, then, writing Chinese characters allowed Branford to be simultaneously foreign and familiar: "He's from abroad, yes, but you know there's a connection there, something strong, some of it words," Okito said. "They always thought he was American—Chinese-American, yes—but when he writes [the *kanji*], it goes to a different dimension you see." Thus, three rhetorical strands work together:

1. Physical appearance, which marks Branford as familiar
2. Spoken language in English, which marks him as foreign; and
3. Competency with written language, which invents a unique identity for him as both foreign and familiar.

According to Okito, this identity creates a rhetorically transformative power: "Before Branford wrote the Chinese characters, when he said something, some people might have got offended. The same person after he wrote that might say, 'Maybe it's me that's not thinking hard enough. . . . Let me pause a little bit and try to see what he's trying to say.'"

This illustrative example is just one of many in the case study where the participants used linguistic competence as a rhetorical strategy within a particular context to build and maintain relationships. Implications of these findings for technical communicators include the following:

- Linguistic competence can have great rhetorical value for international professional communicators who wish to build ethos with their audiences.
- The value of *limited* linguistic competence to build rapport should not be ignored or underestimated.
- To realize the rhetorical benefits from using linguistic competence, technical communicators must be aware of cultural expectations that constrain or enable their use of the language.

These findings and related implications are just a new iteration of social-rhetorical theories widely accepted in the field of technical communication. For example, Berlin's (1987) social-epistemic theory of rhetoric asserts that communities are built around language peculiar to themselves, and membership in a group is determined by the ability to use language competently (pp. 166–167).

Most technical communicators would likely agree with these statements and reject the notion that rhetoric is merely a device for transferring information cleanly (Bazerman, 2004; Selzer, 2003; Winsor, 2003). However, gaining foreign language competence can be such a daunting task that technical communicators may focus only on the practice of "getting it right," or, alternately, not attempt to learn or speak a foreign language at all. They may overlook the *rhetorical* aspects of speaking a language, including opportunities to build rapport through limited linguistic competence. Not recognizing these opportunities, they may also fail to consider cultural expectations that shape the way they are perceived as speakers of a language, either constraining or enabling certain kinds of language use.

For technical communication programs and students, the implications are several. Students who gain a nuanced view of communication in a second language will be aware of the role of second-language abilities in global contexts and will perhaps be encouraged to pursue the acquisition of a second language, even if the prospects for gaining full fluency seems unlikely. Technical communication instructors can focus more on second-language issues from the perspectives of L1 and L2 English speakers, both in standard courses and in courses specifically devoted to this purpose. Programs can facilitate this process by arranging study abroad or international internships, by focusing more language-related issues in technical communication class, and by working to collaborate on curriculum with foreign language departments in the academy.

Translation

I expected translation to be an important area of competence for the Japanese-speaking team members because they do most of the written translation as well as the real-time oral interpretation of the training presentation. (In this chapter, I use "translation" as a default term for both written translation and oral interpretation.) However, the whole process of translation turned out to be much more rhetorically grounded, collaborative, and relationship-oriented than I had imagined and required each member of the team, whether translator or translatee, to be competent. I found that the quality and efficiency of the team's translation was enhanced by

1. Team member-translators (Okito and Emi) who were familiar with the context and audience of the training presentation,
2. The long-term relationship between translator and translatee,
3. A team member, Okito, who was bicultural in addition to being bilingual, and
4. Team members who were able to collaborate in both the "big picture" process and the technical aspects of translation and interpretation.

First, a key to Okito's and Emi's effectiveness as translators was their familiarity with the context of the training—including the industry, company, and product—a familiarity the participants stressed had developed over a period of 5 years. In addition to allowing for a more precise, contextually based vocabulary to develop, participants said this grounding was necessary to understand the overall direction and purpose of the training.

At the team's beginning, Branford, who had been a professional trainer for about 20 years and was already familiar with the company and product, presented the training content, and Okito and Emi acted primarily as on-the-spot, rote interpreters. Over time, these roles shifted, as the participants explained in a group interview:

BRANFORD: Now remember, they [Okito and Emi] didn't know what they were doing [at the beginning], so they were hearing a lot of the stuff for the first time.

OKITO: Yeah, that's true.

BRANFORD: So we just went to Japan and went at it slowly. Between what we did there and what we did here [in Hawaii], they started to understand all of the training. That was a big breakthrough because they could anticipate the direction that we were going in and knew the information.

Second, over the period of 5 years, the three team members had also built a relationship with one another. A recurring theme in my conversations with the participants was the distinction between being able to translate words and being able interpret thoughts, feelings, ideas, and intent. Okito explained, "Giving them a translation of what Branford says is not enough. You've got to become Branford."

At the time I observed the team, Emi and Okito had become active contributors throughout the creation and delivery of training. Because of the integration of their roles as translators and trainers, they could represent not only the message but the messenger; indeed, they were messengers themselves. The familiarity they had developed with the training content, audience, and each other allowed them to collaborate fully and actively not only on translation, but also on content invention and delivery.

Hoft (1995) recognizes the importance of a translator's familiarity with both context and collaborators, noting that "many companies now invest in training translators, even if the translators work for a translation company" (p. 195). The purpose of this training is usually to introduce translators to the project team and help them learn about the products. In the case study, Okito and Emi were already part of the project team, so they had gained a built-in, on-the-job familiarity with each other and with the training content and context.

A third important factor was something that all three participants noted: having Okito as a bicultural member of the team greatly increased the quality of translation. Emi, who was uniquely qualified to evaluate this ability because she had previously operated her own translation company, said Okito was an excellent translator because, in addition to speaking both languages, "he knows both cultures [and] both mentalities," and so was able to translate words with full comprehension of cultural context. Branford added, "I could be a terrific Japanese speaker, but unless I know Japanese culture I'm not going to be effective. That's why the key is someone like Okito if you're working in a foreign market." Similarly, good international technical communication practice requires expert advice on both the linguistic and cultural levels. For this reason, technical communicators who can translate or collaborate effectively with translators will be highly valued.

As an expert on both the company and cultural contexts, Okito was able to take an informed, rhetorical approach to translation. While much of the translation of training content such as PowerPoint slides was, in Okito's words, "just straight translation," some phrases were modified to fit the cultural values of the audience. For example, at one point in the English version of the training, the presenters talk about focus and why it is necessary for success in sales. Okito explained that he and Emi had instead chosen to use a Japanese phrase that, translated directly, means "drench head," with a connotation of being deeply devoted and immersed. In English, such a phrase might be seen as too strong. However, Okito and Emi believed this emphasis would be appropriately received by the audience members in Japan. Similar kinds of adaptations were made in other parts of the presentation. Again, relationship was important because the team members trusted Okito and were able to draw on his bicultural expertise, resulting in excellent translation. This finding is applicable to all international technical communicators—better relationships with team members who are cultural insiders will facilitate better translation and cultural adaptation.

A fourth factor in effective translation was that, in addition to having contextual familiarity and cultural fluency, team members were able to collaborate in both the "big picture" process and the technical aspects of translation and interpretation. Okito and Emi often consulted with each other and with other Japanese speakers to get a consensus opinion on the best translation for the slides. Branford, although not a translator, had a sophisticated awareness of the translation process. He also used subtle but essential competencies as a translatee, including technical aspects of working with interpreters, such as adequate pausing, as well as understanding the overall process of translation and how he fit into it. Each individual was able to contribute his or her individual abilities toward translation, resulting in effective collaboration for the team. Similarly, all international technical communicators need to have a solid understanding of the translation process and of technical aspects, such as pausing, of working effectively with translators.

These findings should not be surprising to technical communicators trained in rhetoric. If language and rhetoric are social, then translation is not merely a mechanical or informational process but a social, rhetorical, and collaborative process connected to social and cultural values. Yet, as technical communicators, we may forget our own foundations when approaching translation in international training. We may harbor concepts of literal translation rather than valuing appropriateness or the overall process of translation (Flint, van Slyke, Starke-Meyerring, & Thompson, 1999; Hoft, 1995; Walmer, 1999). As with the writing process, translation of international training content often requires not only the observance of certain prescribed conventions, but also requires a much broader negotiation of social and cultural expectations and values to create shared knowledge (Flint et al., 1999; Walmer, 1999).

Technical communicators will have an advantage in international training if they approach translation not just as a mechanical process but as a rhetorical and relationship-oriented process, complete with audience, author, and message, all filtered through cultural values and expectations (Bazerman, 2004; Cross, 2001; Winsor, 2003). In fact, the rhetorical dimension of written, oral, and written-oral hybrid professional communication may become even more apparent to international trainers as they are forced to choose a translation strategy (Hoft, 1995), build relationships with translators (Walmer, 1999), and view their own expectations and values about communication in light of other cultures (Connor, 1996; Thatcher, 2000).

Similar to the implications of findings related to second-language competence, technical communication programs need to facilitate a better understanding of the translation process and teach students how to effectively work with translators (Maylath, 1997). Instructors and students can make connections between translation theory and the rhetorical theories they are more familiar with. They might collaborate with foreign language departments on translation projects. Even better, they might link up with translation students in another part of the world, as is being done in a "Trans-Atlantic Collaboration between U.S. Instructors of Technical Writing and EU Instructors of Translation" (Humbley, Maylath, Mousten, Vandepitte, & Veisblat, under review). As with efforts to acquire greater L2 abilities, international internships can also play an invaluable role as a capstone for the theory and hands-on experience students gain.

Intercultural Communication

Intercultural communication competence is the ability to negotiate divergent cultural values and their corresponding influences on communication (Byram, 1997; Hampden-Turner & Trompenaars, 2001). Participants used intercultural competence in two ways: (a) adapting the training seminar to fit the cultural values and expectations of the Japanese's audience, and (b) managing the context surrounding the training seminar to build rapport with the audience.

To evaluate the participants' rhetorical adaptation of the seminar, I compared the texts of two PowerPoint training presentations the participants made, one a presentation in Hawaii for a mixed audience of Americans and Japanese in both English and Japanese, and one in Japan for a Japanese audience only. I found that the participants made two main rhetorical adaptations between the presentations: first, they used more detail in the Japanese version; and second, they used graphics more heavily in the Japanese version.

I had expected the texts of the PowerPoint presentations to differ according to the theory of contrastive rhetoric (Connor, 1996; Kaplan, 1966, 1987), in which arrangement would cater to the language and cultural background of the audience.

For the Japanese audience, I expected a more delayed introduction of purpose, or quasi-inductive arrangement (Hinds, 1990) than for the hybrid audience. However, I was surprised to find that the arrangement of both presentations was quite similar. The major difference was that the Japanese seminar contained more information on sales strategies (16 slides versus only 12 slides for the hybrid presentation) and a 10-minute role-play sales demonstration in Japan versus none in Hawaii. When asked why, the participants responded that Japanese audience members generally want more detail than Americans, who are often content with a broad overview (Yamada, 1997).

The use of graphics was also heavier in the Japan seminar, although information in both seminars was depicted visually, with several slides using standalone graphics, including cartoons, to illustrate the presenters' points. One indicator of the difference in use of graphics was that the Japanese presentation had 10 graphics-only slides, while the hybrid seminar had just 5. When asked about this difference, Okito and Emi said they were catering to the Japanese preference for visual over written information (Kohl, Barclay, Pinelli, Keene, & Kennedy, 1993; Yamada, 1997). They added that graphics tend to lighten the mood of the presentation in Japan, which might become too serious otherwise: "We try to insert some [graphics] in to make it not so intense but understandable." Participants were aware of these differing audience expectations about the text and responded accordingly.

Perhaps more importantly, the participants used an intercultural approach to manage the context surrounding the training seminar. They were highly sensitive to what Hall (1976) defines as high-context communication, or communication that comes primarily through the context of a message rather than the content. For example, all of the Japan meetings included a dinner or banquet afterwards, which the participants said was standard practice for the team there, but not in the United States. The participants said these functions were more important than the training seminar itself because in an informal setting, audience members could ask questions or raise concerns that they would be reluctant to mention in the meetings (Hall, 1990; Yamada, 1997). Okito said the tendency of Japanese audience members to ask questions only in an informal setting stemmed either from the concern that they might lose face by asking an unintelligent question in front of everyone or that they might make the presenter lose face by asking a question he or she couldn't answer (Hampden-Turner & Trompenaars, 2000; Masumoto, Oetzel, Takai, Ting-Toomey, & Yokochi, 2000). The fact that the training team participated in a social event after the training session was an unspoken message that they were personally committed to helping the audience members become better salespeople.

Beyond the expertise required to recognize the importance of social events, the posttraining banquets and dinners were themselves infused with nonverbal communication, requiring cultural expertise on the part of the team members. For example, it was important that people be seated in their proper places at the table; this nonverbal cue, defined according to a culturally defined hierarchy, maintained the cohesion of the group and helped everyone feel there was a proper place for them, both at the table and in the organization (Hall, 1990; Yamada, 1997). The Japanese team members were able to navigate these kinds of cultural protocols seemingly without effort. The American team member was also able to negotiate these events, an ability the research participants said he had gained through experience and by seeking feedback from others, particularly his teammates.

Overall, team members adapted not only the content and delivery of the presentation for their Japanese audience, but also took an intercultural approach to the surrounding events, treating the entire training effort as an integrated whole. In these efforts, the team members emphasized their collaboration with each other and with their associates in Japan. In an interview a few days after the training seminar in Japan, Okito stated, "I think our work really paid off, because we didn't just whip something out and just bring it in here . . . [but] really we discussed it and talked our situation [through] with people [so] that we feel it is timely and really worked on it together." He added, "I think, overall, through the long process of five years, we're to the point where we know the key elements to bring up, and I think that's what it is."

Implications of these findings for technical communicators include the following:

In international training, communicative expertise cannot be limited to just speaking or writing (Spilka, 1993; Thatcher, 1999). Trainers must recognize what Hall (1976) calls "the total communication framework," which includes "words, actions, postures, gestures, tones of voice, facial expressions, the way [a person] handles time, space, and materials, and the way he works, plays, makes love, and defends himself." As Hall explains, "All these things and more are complete communication systems with meanings that can be read correctly only if one is familiar with the behavior in its historical, social, and cultural context" (p. 42).

Because of the complexity of intercultural communication, extensive personal and team preparation is essential (Hoft, 1995). International trainers can't assume that international audiences will respond in the same way as domestic audiences. As a first step, they might develop "localization cookbooks," which provide "cultural, linguistic, technical, marketing, sales, training, and legal information about a specific target country and the users who live and work there" (Hoft, 1995). Even more important, international trainers should not overlook the necessity of building strong relationships with cultural insiders who can help them adapt to situational nuances. They should not forget that they can compensate for competency gaps by collaborating with others, an ability that is discussed in the next section.

Implications for students and instructors are many. Professional communication programs that assume common American communication practices as universal may fail to prepare students for international professional communication contexts. Building the ability to reconcile cultural differences in the areas of context management, relationship building, and high-context communication should be important goals of intercultural curriculum, in addition to a focus on document creation and project management. The value of exchanges and other experience-based learning should not be underestimated for as E. T. Hall (1989) notes, "Most cross-cultural exploration begins with the annoyance of being 'lost,'" a condition that requires real intercultural interaction. In teaching, models of intercultural professional communication competence should include relationship building as a key purpose and goal of communication.

COLLABORATION

Collaborative competence refers to the ability of individuals to adapt to new environments in collaboration with team members. In the case study, if team members could collaborate with one another, it wasn't necessary for each individual to have all of the competencies necessary. However, this collaboration did require the participants to have two main abilities:

1. Giving and receiving feedback,
2. Recognizing and using individual competencies to compensate for gaps in team members' abilities.

With regard to the first ability, feedback was given both by team members and by outside associates. During team meetings, Okito and Emi gave feedback on what would be culturally appropriate. For example, Branford wanted the audience of salespeople to fill out a worksheet on financial goals but Okito and Emi were skittish about it:

> BRANFORD: I got a work sheet for them where they can actually be involved.
>
> OKITO: Do you have that worksheet [with you]?
>
> BRANFORD: Um, I didn't bring it with me. I was just gonna, you know, savings you know, federal retirement, what you need to retire, and . . . take 'em through each of the steps.
>
> OKITO & EMI: [speaking to each other in Japanese]
>
> OKITO: Yeah we can't let it because that won't be appropriate.

Okito later explained that their concern was based on Japanese cultural values that asking people to write down information involving personal finances in a meeting, especially if they thought they would have to show their goals to a supervisor, might cause them to lose face (Masumoto et al., 2000). Branford still wanted to include the sheet in the meeting, reasoning, "You need to get them somehow thinking about it. If they don't think about it, they don't do something." He added that the sheet wouldn't have to be completed at the meeting but could be taken home. The team resolved the dilemma by passing out the worksheet in the meeting and making it clear that if the audience members wanted to, they could take it home and set voluntary goals. Audience members also had the option of showing their goals to a supervisor but did not have to.

Similar reconciliations between cultural values were made in other planning meetings. Branford explained, "I understand [some things], but there are other things that are very hard for me to understand, so sometimes [Emi and Okito] spend hours trying to help me understand. If you understand the situation then you can correct [your errors.]" At the same time, participants recognized the value of an outside perspective. Okito said, "Sometimes the unusual happens too. Sometimes we talk to Branford, we might take an hour or so, [and] Branford will come up with this totally Western version of what a Japanese might be able to do, which becomes very good because it's a new idea."

Team members also sought feedback from associates in Japan, both before and after the training. Getting this feedback required intercultural expertise beyond the ability to speak Japanese. Okito and Emi stated that because they were geographically separated from their Japanese associates, they often had to encourage the use of e-mail for questions or concerns. In Japan, the more important the communication is, the more likely it will be face-to-face (Yamada, 1997). Thus, there is often a tendency to put off discussion until a face-to-face meeting can be arranged. Since the training team only traveled to Japan three or four times a year, a face-to-face meeting might not happen for months. Not only would it be more difficult to resolve questions and concerns at such infrequent intervals, but it would be difficult to create training that met real needs. As Emi noted, "To know the feedback from people after the training is very important, sometimes more important than the training."

Thus, a subtle but vital part of the training Okito and Emi did was encouraging Japanese associates to use remote communications technology in ways they were not accustomed to. With regard to this effort, the participants emphasized the long-term nature of their relationship with associates in Japan. Over time, trust had become strong enough to support more remote communication, and associates felt comfortable sharing questions and concerns through phone or e-mail rather than reserving them for face-to-face communication. As with translation competence, giving and receiving useful feedback both within and outside the team was enhanced by long-term relationships.

A second important collaborative ability was using one's individual competencies to compensate for gaps in team members' competencies, and allowing them to do the same. This chapter has already reviewed how team members combined their individual competencies in the areas of linguistic and translation competence. In another example, team members collaborated to reconcile culturally defined expectations about age and gender. Referring to himself and Emi, Okito noted, "Certain things we cannot say that Branford can say," and cited an example: "I cannot tell a guy in his sixties 'you'd better prepare [for] retirement.'" Such a statement would be offensive coming from a man in his late thirties or early forties, given Japanese the cultural value of respect for elders (Yamada, 1997). Okito added that "certain directions have more weight" when coming from someone with "higher seniority and more status," which in Japan is often defined by age (Yamada, 1997). He noted, "In some cases, we repeat it too. Certain people—younger people—might relate more to us, middle-aged people might relate to Branford and us, but the older generation might only respect what Branford says."

Another important collaboration between team members involved gender roles. Traditional Japanese gender roles limit how women are able to act in relation to men in the business world (Hall, 1990; Yamada, 1997). However, this limitation also poses a barrier to men who are in positions of authority; women are less likely to voice opinions or concerns (Hall, 1990; Yamada, 1997). Since over half of the salespeople in the organization were women, Emi's role as listener and confidant was important. It was to her that much of the feedback from female personnel was directed. Emi met with female salespeople individually, and during group events, prepared the way for more open interaction between the male training team members and female salespeople.

These examples illustrate how the team acted as an integrated whole, requiring each member to recognize strengths and weaknesses and rely on others' unique abilities. In this way, the team could accomplish things none of the participants could do alone. Each had to be willing to draw upon and contribute strengths throughout the process (Larson & LaFasto, 2001). Participants were careful to note that building their relationship as a team was at least as important as developing the content of the training. On several occasions team members reminded me that they had been working with each other and some of their associates in Japan for 5 years. They saw this process and the training itself as a long adaptation—an evolving, "living thing." Team members needed to be familiar with the way each other thought, especially since Okito and Emi would be interpreting Branford's portion of the seminar. They emphasized that developing such a connection took time and often happened in indirect ways.

Implications for international trainers include the following:

- Collaborative competence is important in any team situation, but in intercultural context requires an even stronger and more long-term collaboration among team members.
- Giving and receiving feedback is key to successful intercultural collaboration.

- Cultural outsiders need to rely on teammates for cultural expertise; at the same time, outsiders may add a valuable perspective to the team.
- Intercultural collaboration abilities are required for collaboration with diverse team members and outside associates. International trainers can adapt to the cultural expectations of others; they can also help others reconcile their cultural values and expectations with the demands of the situation (Hampden-Turner & Trompenaars, 2004). As the e-mail example demonstrates, intercultural collaboration is intercultural for all collaborators, not just for the "sojourner." (Byram, 1997)

While collaboration has long been recognized as vital to successful technical communication projects (Dicks, 2004; Hackos, 1994), technical communicators must be flexible about what collaboration means in different international training contexts. As cultural expectations vary, so does the cultural expression of collaboration. Hierarchy versus egalitarianism, high-context versus low-context communication, individualism versus collectivism, and gender roles are all areas where cultural expectations about collaboration might diverge (Hall, 1976; Hampden-Turner & Trompenaars, 2004; Hofstede, 1996). By integrating intercultural communication competence and collaborative competence, technical communicators can work successfully with training-team members and audiences from diverse cultural backgrounds.

Instructors and students need to recognize that collaboration practices are not necessarily universal, and those that seem "natural" are likely rooted in North American cultural values and assumptions (Bosley, 2004). Because collaboration in many other cultural contexts such as Japan tends to be more interdependent and less goal-oriented than in the United States, professional communication students need to learn a flexible approach to collaboration. Success in all other areas of competence will depend on the ability to collaborate globally. Thus, in addition to an encouraging increased classroom focus on collaboration, technical communication programs should facilitate international internships, exchanges, and global educational partnerships such as the translation-focused Transatlantic Project (Humbley et al., under review) and the Global Classroom Project (Herrington & Tretyakov, under review; Kennon, 2008).

CONCLUSION

The case study discussed in this chapter shows that revising assumptions about the primary purpose of professional communication from information sharing to relationship building also results in a revision of all related competencies. Language fluency, rather than facilitating the efficient transfer of information, can build bridges between speaker and audience, even when actual language competence is weak. Translation ability is both a result of and an enabler of strong relationships among team members and between audience and presenters. Intercultural communication competence also results from and builds these strong relationships. Effective collaboration in all of these areas is, in Okito's words, like a "living thing" that grows during a larger overall process. Such changes in the way international professional communication competencies are viewed result from revising just one common assumption about information sharing and relationship building. Other revisions might focus on common assumptions about the feasibility of addressing culture without also addressing language, the perceived worldwide predominance of American English (Melton, under review), and the valorization of writing over other kinds of communication (Thatcher 1999, 2000; Yamada, 1997).

For teachers and students, implicit in such an effort is an ability that can be added as an umbrella competence: adaptation. Also called fundamental competence (Spitzberg & Cupach, 1984), adaptive competence is the ability to adapt to meet changing conditions. In sum, it is "knowing what you don't know" and how to get it. International training team members (and the programs and instructors who prepare them) must be able to recognize what skills and knowledge are needed, discover ways to obtain them, decide when to collaborate, assess the success of their communication strategies, and adjust accordingly. Technical communication practitioners and instructors often approach problems in terms of adaptation (audience analysis and usability testing being two examples of this tendency) and can apply these abilities to international training situations. However, adaptation that is useful in a domestic context may find different expression in an intercultural context, requiring a different approach than technical communicators might be used to. Cultures may have differing views of such basic concepts as time, leadership, collaboration, and communication, requiring an openness to new learning and ways of doing things (Hoft, 1995).

Ways for technical communication students and teachers to gain perspective on adaptation might include foreign language learning (Byram, 1997), international internships (Smith, 2003), classroom instruction with intercultural focus and international interaction (Connor, Davis, Derycker, Philips, & Verkens, 1997; Sapp, 2004), and global educational partnerships (Kennon, 2008). Through such efforts, the field of technical and professional communication can make any necessary revisions to long-held assumptions and "[address] basic yet also advanced skills that are [still often] hidden by our very definition of the professional communicator and the principles of professional communication" (Weiss, 1997, p. 336).

REFERENCES

Anderson, P. (2002). *Technical communication: A reader-centered approach.* Boston: Heinle.

Armstrong, A. (2002). Applying instructional design principles and adult learning theory in the development of training for business and industry. In P. Rogers (Ed.),

Designing instruction for technology-enhanced learning (pp. 209–227). Hershey, PA: Idea Group Publishing.

Bazerman, C. (2004). Speech acts, genres, and activity systems. In C. Bazerman & M. Prior (Eds.), *What writing does and how it does it* (pp. 309–339). Mahwah, NJ: Erlbaum.

Bennett, M. (1998). Overcoming the golden rule: Sympathy and empathy. In M. Bennett (Ed.), *Basic concepts of intercultural communication*. Yarmouth, ME: Intercultural Press.

Berlin, J. A. (1987). *Rhetoric and reality: Writing instruction in American colleges, 1900–1985*. Urbana, IL: CCCC.

Bosley, D. (2004). Cross-cultural collaboration: Whose culture is it, anyway? In S. Selber & J. Johnson-Eilola (Eds.), *Central works in technical communication* (pp. 466–474), New York: Oxford University Press.

Bridges, B., Baily, M., Hiatt, M., Timmerman, D., & Gibson, S. (2002). A blended technologies learning community: From theory to practice. In P. Rogers (Ed.), *Designing instruction for technology-enhanced learning* (pp. 209–227). Hershey, PA: Idea Group.

Burba, F. J., Petrosko, J. M., & Boyle, M. A. (2001). Appropriate and inappropriate instructional behaviors for international training. *Human Resource Development Quarterly, 12,* 267–283.

Buckman, R. H. (2004). *Building a knowledge-driven organization*. New York: McGraw-Hill.

Byram, M., (1997). *Teaching and assessing intercultural competence*. Cambridge, UK: Multilingual Matters.

Campbell, C. P. (1998). Rhetorical ethos: A bridge between high-context and low-context cultures? In S. Niemeier, C. P. Campbell, & R. Dirven (Eds.), *The cultural context in business communication* (pp. 31–47). Philadelphia: John Benjamins.

Carliner, S. (2001). Emerging skills in technical communication: The information designer's place in a new career path for technical communicators. *Technical Communication, 48*(2), 156–175.

Chen, G.-M. (1989). Relationships of dimensions of intercultural communication competence. *Communication Quarterly, 37,* 118–133.

Chen, G.-M., & Starosta, W. J. (1996). Intercultural communication competence: A synthesis. *Communication Yearbook, 19,* 353–383.

Connor, U. (1996). *Contrastive rhetoric: Cross-cultural aspects of second-language writing*. Cambridge, UK: Cambridge University Press.

Connor, U., Davis, K. W., De Rycker, T., Philips, E. M., & Verkens, J. P. (1997). An international course in international business writing: Belgium, Finland, and the United States. *Business Communication Quarterly, 60*(4), 63–74.

Cross, Geoffrey, A. (2001). *Forming the collective mind: A contextual exploration of large-scale collaborative writing in industry*. Cresskill, NJ: Hampton Press.

Crystal, D. (2003). *The Cambridge encyclopedia of the English language* (2nd ed). Cambridge, UK: Cambridge University Press.

Dicks, R. S. (2004). *Management principles and practices for technical communicators*. New York: Pearson.

Earley, P. C., & Gibson, C. B. (2002). *Multinational work teams*. Mahwah, NJ: Erlbaum.

Faber, B., & Johnson-Eilola, J. (2002). Migrations: Strategic thinking about the future(s) of technical communication. In B. Mirel & R. Spilka (Eds.), *Reshaping technical communication: New directions and challenges for the 21st century* (pp. 135–148). Mahwah, NJ: Lawrence Erlbaum.

Flammia, M. (2005). Preparing technical communication students to play a role on the translation team. *IEEE Transactions on Professional Communication, 48*(4), 401-412.

Flick, U. (2002). *An introduction to qualitative research*. London: Sage.

Flint, P., Van Slyke, M. L., Starke-Meyerring, D., & Thompson, A. (1999). Going online: Helping technical communicators help translators. *Technical Communication, 46,* 248–248.

Hackos, J. (1994). *Managing your documentation projects*. New York: Wiley.

Hager, P. (2000). Global graphics: Effectively managing visual rhetoric for international audiences. In P. Hager & H. Scheiber (Eds.), *Managing global communication in science and technology*. New York: John Wiley & Sons.

Hall, E. T. (1976). *Beyond culture*. Garden City, NY: Anchor.

Hall, M. (1990). *Hidden differences: Doing business with the Japanese.* Garden City, NY: Anchor.

Hampden-Turner, C., & Trompenaars, A. (2000). *Building cross-cultural competence.* New York: John Wiley and Sons.

Hampden-Turner, C., & Trompenaars, A. (2004). *Managing people across cultures.* West Sussex, UK: Capstone.

Hart-Davidson, W. (2001). On writing, technical communication, and information technology: The core competencies of technical communication. *Technical Communication, 48,* 145–155.

Herrington, T., & Tretyakov, Y. (under review). Processes of partnering: Building the partnering keystone for global classroom project construction and maintenance. In D. Starke-Meyerring & A. H. Duin (Eds.), *Designing global learning environments: Visionary partnerships, policies, and pedagogies.* Rotterdam, Netherlands: Sense Publishers.

Hinds, J. (1990). Inductive, deductive, quasi-inductive: Expository writing in Japanese, Korean, Chinese and Thai. In U. Connor & A. Johns (Eds.), *Coherence in writing: Research and pedagogical perspectives* (pp. 91–110), Alexandria, VA: TESOL.

Hofstede, G. (1996). *Cultures and organizations: Software of the mind.* New York: McGraw-Hill.

Hoft, N. (1995). *International technical communication: How to export information about high technology*. New York: Wiley.

Johnson-Sheehan, R. (2005). *Technical communication today*. New York: Longman.

Kaplan, R. B. (1966). Cultural thought patterns in intercultural education. *Language Learning, 16*, 1–20.

Kaplan, R. B. (1987). Cultural thought patterns revisited. In U. Connor & R. B. Kaplan (Eds.), *Writing across languages: Analysis of L2 text.* Reading, MA: Addison-Wesley.

Kennon, J. (2008). Learner perspectives: International collaboration and cross-cultural communication in the Global Classroom Project. In D. Starke-Meyerring & A. H. Duin (Eds.), *Designing global learning environments: Visionary partnerships, policies, and pedagogies* (pp. 114-128). Rotterdam, Netherlands: Sense Publishers.

Kohl, J., Barclay, R., Pinelli, T., Keene, M., & Kennedy, J. (1993). The impact of language and culture on technical communication in Japan. *Technical Communication, 40,* 62–73.

Larson, C., & LaFasto, F. (2001). *When teams work best.* Thousand Oaks, CA: Sage.

Lazear, E. P. (1998). *Globalization and the market for teammates.* Working Paper Series: National Bureau of Economic Research. Cambridge, MA: NBER.

Locker, K., & Kaczmarek, S. (2003). *Business communication: Building critical skills.* New York: McGraw-Hill/Irwin.

Lucia, A. D., & Lepsinger, R. (1999). *The art and science of competency models: Pinpointing critical success factors in organizations.* San Francisco, CA: Pfeiffer.

MacKinnon, J. (1993). Becoming a rhetor: Developing writing ability in a mature, writing-intensive organization. In R. Spilka (Ed.), *Writing in the workplace: New research perspectives.* Carbondale, IL: Southern Illinois University Press.

Marquardt, M. (1995, May). *How to globalize your training.* INFO-LINE. Alexandria, VA: ASTD.

Masumoto, T., Oetzel, J. G., Takai, J., Ting-Toomey, S., & Yokochi, Y. (2000). A typology of facework behaviors in conflicts with best friends and relative strangers. *Communication Quarterly, 4*(48), 397.

Maylath, B. (1997). Writing globally: Teaching the technical writing student to prepare documents for translation. *Journal of Business and Technical Communication, 11,* 339–352.

McDermott, L., Waite, B., & Brawley, N. (1999). Putting together a world-class team. *Training & Development, 53*(1), 46.

Melton, J. H. (2008a). Beyond standard American English: (Re)thinking language in globally networked learning environments. In D. Starke-Meyerring & A. H. Duin (Eds.), *Designing global learning eEnvironments: Visionary partnerships, policies, and pedagogies* (pp. 185-199). Rotterdam, Netherlands: Sense Publishers.

Melton, J. H. (2008b). Language, culture, and collaboration in offshore outsourcing: A case study of international training team communication competencies. In C. Evia & B. Thatcher (Eds.), *Outsourcing technical communications: Policies, practices, and procedures* (pp. 123-145). Amityville, NY: Baywood.

Miles, M., & Huberman, M. (1994). *Qualitative data analysis: An expanded sourcebook.* Thousand Oaks, CA: Sage.

Mirel, B., & Spilka, R. (2002). Introduction. In B. Mirel & R. Spilka (Eds.), *Reshaping technical communication: New directions and challenges for the 21st century.* Mahwah, NJ: Lawrence Erlbaum.

Olebe, M., & Koester, J. (1989). Exploring the cross-cultural equivalence of the behavioural assessment scale for intercultural communication. *International Journal of Intercultural Relations, 13,* 333–347.

Prahalad, C. K. (1997). The work of the new age managers in the emerging competitive landscape. In F. Hesselbein, M. Goldsmith, & R. Beckhard (Eds.), *The organization of the future.* San Francisco, CA: Jossey-Bass.

Rainey, K., Turner, R., & Dayton, D. (2004, October 7). Report of a survey of managers about core competencies. Presented at the annual meeting of the Council for Programs in Technical and Scientific Communication, West Lafayette, IN. Retrieved from http://www.english.vt.edu/~dubinsky/CPTSC_04/program_links.htm

Ramsey, S. J. (1998). Interactions between North Americans and Japanese: Considerations of communication style. In M. Bennett (Ed.), *Basic concepts of intercultural communication*. Yarmouth, ME: Intercultural Press, Inc.

Reich, R. B. (2001). *The future of success*. New York: Knopf.

Richmond, Y. (1995). *From da to yes: Understanding east Europeans*. Yarmouth, ME: Intercultural Press, Inc.

Ruben, B. D. (1976). Assessing communication competency for intercultural adaptation. *Group and Organization Studies, 2,* 470–479.

Ruben, B. D., & Kealey, D. J. (1979). Behavioral assessment of communication competency and the prediction of cross-cultural adaptation. *International Journal of Intercultural Relations, 3,* 15–47.

Sapp, D. A. (2004). Global partnerships in business communication: An institutional collaboration between the United States and Cuba. *Business Communication Quarterly, 67,* 267–280.

Schwartz, S., & Sagiv, L. (1995). Identifying culture-specifics in the content and structure of values. *Journal of Cross-Cultural Psychology, 26,* 92–116.

Selzer, J. (2003). Rhetorical analysis: Understanding how texts persuade readers. In C. Bazerman & M. Prior (Eds.), *What writing does and how it does it: An introduction to analyzing texts and textual practices* (pp. 279–308). Mahwah, NJ: Erlbaum.

Smith, H. J. (2003). German academic programs in technical communication. *Journal of Technical Writing & Communication, 33,* 349–363.

Spencer, L. M., & Spencer, S. M. (1993). *Competence at work: Models for superior performance*. New York: Wiley.

Spilka, R. (1993). Moving between oral and written discourse to fulfill rhetorical and social goals. In R. Spilka (Ed.), *Writing in the workplace: New research perspectives*. Carbondale, IL: Southern Illinois University Press.

Spitzberg, B. H., & Cupach, W. R. (1984) *Interpersonal communication competence*. Beverly Hills, CA: Sage.

Thatcher, B. (1999). Cultural and rhetorical adaptations for South American audiences. *Technical Communication, 46,* 177–195.

Thatcher, B. (2000). Writing policies and procedures in a U.S. and South American context. *Technical Communication Quarterly, 94,* 365–400.

Thornhill, A. R. (1993). Management training across cultures: The challenge for trainers. *Journal of European Industrial Training, 17*(10), 43–51.

Walmer, D. (1999). One company's efforts to improve translation and localization. *Technical Communication, 46,* 230–237.

Weiss, T. (1997). Reading culture: Professional communication as translation. *Journal of Business and Technical Communication, 11,* 321–338.

Weiss, T. (1999). The implications of translation for professional communication. In C. Lovitt & D. Goswami (Eds.), *Exploring the rhetoric of international professional communication: An agenda for teachers and researchers* (pp. 277-310). Amityville, NY: Baywood.

Whiteside, A. L. (2003). The skills that technical communicators need: An investigation of technical communication graduates, managers, and curricula. *Journal of Technical Writing and Communication, 33,* 303–318.

Winsor, D. A. (1996). *Writing like an engineer: A rhetorical education*. Mahwah, NJ: Lawrence Erlbaum.

Winsor, D. A. (2003). *Writing power: Communication in an engineering center*. Ithaca, NY: SUNY Press.

Yamada, H. (1997). *Different games, different rules: Why Americans and Japanese misunderstand each other*. Oxford, UK: Oxford University Press.

Yin, R. (1994). *Case study research: Design and methods*. Thousand Oaks, CA: Sage.

Yuan, R. (1997). Yin/yang principle and the relevance of externalism and paralogic rhetoric to intercultural communication. *Journal of Business & Technical Communication, 11*, 297–320.

CHAPTER 8

Teaching Technical Communication in France: Challenges and Prospects

Dacia Dressen-Hammouda

The importance of teaching written communication is a well-established tradition in both the United States and Canada, in parts of South America, as well as throughout northern Europe, including Germany, the Netherlands, Denmark, Scandinavia, and Great Britain. However, while writing instruction in France is heavily emphasized in primary and secondary school education, it has until recently received only very limited attention at the postsecondary (or university and college) level. This situation, of course, has important consequences for potential French degree programs in technical communication since written communication skills are among the most essential for entry into the professional community.

This chapter describes the ongoing implementation of a 2-year master's degree program in intercultural technical communication at a midsized university in central France (Université Blaise Pascal, Clermont 2).[1] The chapter will begin by describing the background context for the curriculum's design, including writing instruction practices at the postsecondary level and a recent reform that allows French universities to begin aligning technical communication programs more closely with North American and northern European practices. After providing a broad overview of the proposed degree program, the chapter will then conclude by briefly discussing implications for developing a curriculum that both raises intercultural awareness and develops necessary written communication skills in French university students.

[1] Also described in Dressen-Hammouda (2006)

BACKGROUND CONTEXT FOR CURRICULUM DESIGN

At the time of this writing, the proposed 2-year master's course in intercultural technical communication was under review by the French Ministry of Education.[2] As with all French university programs, the centralized government retains the final decision on which programs will be accredited and which will not. What this implies is that any university program, because it is largely funded by the French government, reflects the country's current geopolitical considerations to at least some extent. The following sections of this chapter will describe this educational context by viewing reasons why training for technical communicators in France has traditionally emphasized instruction in translation rather than in written communication. It will also provide a brief overview of the types of programs that to date have trained technical writers in France. Such an approach is meant to familiarize the reader with the nature of training in intercultural technical communication in one of the European Union's founding member states.

In effect, what is called "technical writing" in France has long been taught through translation. The university programs that train students to work as technical writers are housed in Applied Foreign Languages for Business and Technology, or L.E.A. (*Langues Etrangères Appliquées*) programs. Historically, such programs specialize not in communication strategies, but in translation techniques (*thème* and *version*). As a result of the consistent emphasis placed on translation, at both the bachelors and the master's level, considerably less attention has been paid to teaching students how to actually *write* the technical and professional documents they will need to produce in the workplace, whether in French or other languages.

While France's geopolitical context explains the heavy emphasis its institutions of higher education place on translation, the preference for translation training also reflects current practices in writing instruction. Notably, university students are expected to have learned the "important" written genres during their secondary school education and to simply be able to adapt their writing skills once they come to the university (Donahue, 2000). As a result, explicit writing instruction is largely absent from the university level in France. Thus, students who train to become future technical communicators often lack experience in writing because they receive little, if any, instruction in writing *non*school (i.e., university and workplace) genres at the university level.

THE FRENCH CONTEXT FOR POSTSECONDARY WRITING INSTRUCTION

The specific context for writing instruction in France thus carries particular implications for designing degree programs in technical communication,

[2] A 4-year accreditation was granted to the program starting in the fall of 2008.

because French students are potentially underprepared for a career in technical communication, with an emphasis on writing.

They do not, however, start out that way. In fact, priority is given to writing instruction throughout the primary and secondary school curricula (Donahue, 2000, 2004). At school, writing is integrated into a whole discoursal approach where children are taught to be aware of language and to use it as a tool in their move toward literacy. This French tradition of "discourse-awareness-rich" (Donahue, 2004, p. 68) teaching begins early in elementary school education, where a writing in the disciplines (WiD) approach underlies the entire design of the curriculum. The WiD emphasis continues throughout secondary school, where writing continues to be essential in every discipline (Donahue, 2004, p. 65). Indeed, primary and secondary students in France have been found to write as much, if not more than, their school-aged American counterparts (Donahue, 2000).

Explanations for French university students' underpreparedness for writing are thus to be found elsewhere. One first explanation can be found in the types of writing instruction students do receive before coming to the university. In effect, writing instruction at the secondary level is geared primarily toward helping students pass the exit exam at the end of high school, called the *Baccalauréat* (commonly referred to as the "Bac"). This institutional need pushes secondary school educators to teach students the types of genres they must master in order to pass the exit exam.[3] Moreover, this educational objective particularly prepares students for one specific type of university writing: end-of-the-semester exam writing, which is also the primary means of evaluation at the university. It is noteworthy that in France such exams are not tied to funding for education, but exist exclusively as a mechanism for evaluating students' performance in a course. Furthermore, given the "massification" of French higher education since the 1980s (Burgel, 2006) and explosion in the number of students coming to the university, such end-of-the-semester exams have become the primary means for evaluation at the university, where individual research and topic papers have become rare (Donahue, 2004).

Another explanation for students' lack of writing experience at the undergraduate level is further tied to the massification of French higher education and the conditions it has created for learning. While some undergraduate degree courses do offer writing classes, called *techniques d'expression*, the results can be quite variable. In effect, during the first 2 years of undergraduate studies in France, class sizes often range upwards to between 150 and 200 students. Such

[3] Such school genres include synthesis essays based on historical or geographical documents, philosophical essays in response to general philosophical essays, or French essays that draw on primary sources and develop themes of analysis: *dissertation, commentaire composé,* etc. (see also Donahue, 2004, p. 65).

conditions make it nearly impossible to engage in "process writing" or to work on controlling the effects of audience, because it is not feasible to provide individual feedback to each student over the course of the semester. As a result, a single writing course is often taught as a large lecture class, and students are invited to work on writing assignments together. Such conditions encourage learning to write as the mastery of stable forms, rather than of the dynamic, rhetorical actions (Miller, 1984) that need to be adapted to the imperatives of each writing situation.

To a certain extent, the lack of emphasis placed on audience-centered writing is also reflected in current French writing research, which tends to focus much more closely on *school* writing than on *university* writing. In effect, the field of rhetoric and composition theory or its equivalent, grounded in postsecondary issues of writing theory and pedagogy, is more or less nonexistent at the French university. In addition, the study of rhetoric, in French called *la stylistique*, is limited to literary courses of study.

Instead, current French writing research reflects the priority placed on teaching writing at the primary and secondary school levels (Donahue, 2000, 2004). To better understand how younger writers become literate, writing researchers have focused on such issues as how

- writing functions as a specific psychological activity (Fayol, 1995, 1996, 1999, 2004; Piolat, 2004; Piolat & Pélissier, 1998);
- children become literate (Altet, 1997; Astolfi, 1993; Bautier, 1995; Garcia-Debanc, 1993; Plane, 2003; Reuter, 1996);
- best to classify and interpret text types considered to be stable forms (Adam, 1992; Haas & Lorrot, 1987); or
- the relationship between language and power is related to success at school (Charlot, Bautier, & Rochex, 1992; Lahire, 1993; Thin, 1998).

As can be surmised from this brief and nonexhaustive list, writing research in France has focused more on the process of literacy acquisition than on describing writing situations and their audience-related concerns.

While another trend in French writing research, from the English for Specific Purposes (ESP) tradition, more closely addresses the concerns of audience, it is only a secondary concern in this research tradition. Here, French university LSP (Languages for Specific Purposes) and ESP teachers and researchers are concerned with the genre-based instruction of specialized discourses for students throughout various disciplines. This trend is represented by two French associations APLIUT (*Association des professeurs de langues des instituts universitaires technologiques*) and GERAS (*Groupe d'études et de recherche en anglais de spécialité*), whose purpose is to contribute to the development of LSP research and teaching in France. While the primary emphasis of research in this domain centers on the teaching of specialized languages in the disciplines, a fair number of these researchers have also worked toward the promotion of writing research and instruction at the postsecondary level in France.[4] It is within this venue that research and teaching practices in France have at times most closely resembled the *rhetorically* centered concerns of technical communication as it is practiced in Anglo-Saxon and northern European countries.

Given the general lack of emphasis placed on writing instruction in the French university curriculum, most students thus gain little experience in writing audience-centered business and professional genres at the undergraduate and graduate levels. The extent to which this may also be true of primary and secondary education is currently the subject of an ongoing investigation. What is clear, however, is that students are largely expected to carry over to the university the writing skills and practices they learned in school (e.g., end-of-the-semester exam writing). These skills are also expected to be carried over to the workplace, despite the difficulties inherent in applying school writing to other types of writing (Beaufort, 1999).

This observation is confirmed by a survey made of 10 existing degree programs in technical translation and writing in France, all but one of which are housed in L.E.A. departments. All such master's-level programs currently known in France were included in the survey. If the university also offered bachelors-level programs, these too were included in the survey. As can be seen in Table 1, the number of hours spent on writing instruction is significantly less than that spent on translation.

As can also be observed in Table 1, most L.E.A. programs that prepare students to work as technical writers prefer *translation*-intensive training over *writing*-intensive training. In effect, students receive only around 40 to 80 hours of writing instruction over the entire course of a 4-year bachelors degree (i.e., between 0.6 and 1.3 hours of writing instruction per week). One can compare these numbers with the number of hours spent on translation training: most degree programs offer around 200 hours of translation in two languages over 4 years (or roughly 9 hours of translation studies per week). Two exceptions to this observation are found at the Université de Bretagne Occidentale, which offers a writing- *and* translation-rich bachelors program, and Université Blaise Pascal. These bachelors programs offer 165 and 128 hours of writing instruction, respectively (or on average, around 2.8 and 2.1 hours per week, respectively).

[4] A special issue of the group's journal, *ASp: la revue du GERAS,* was specifically dedicated to the problem of *rédaction,* or writing research and instruction in France. The researchers who have contributed to this important bed of research have coined the term *rédactologie* in response to the well-established French tradition of the study of translation (*traductologie*). See *ASp*, 37–38 (2002), and notably articles written by S. Birch-Bécass, C. Sionis, K. Nakbi, and R. Cooke, www.geras.fr/dossiers/dossiers. php?val=24_n%BO7+38

Table 1. French Universities with Technical Writing Programs at the Undergraduate and Graduate Levels (Program Survey Conducted June 2006)

Name of university	Master's level	Bachelors ('License') level
Université de Bretagne Occidentale	Master L.E.A.: Rédacteur – Traducteur (given over 1 year) Translation courses: 222 hours Writing courses: 30 hours Content: Unknown	License L.E.A.: Langues Etrangères Appliquées (given over 4 years) Translation courses: 299 hours Writing courses: 165 hours Content: basic writing skills in French, media writing, business writing in English, German/ Spanish, writing longer documents
Université Blaise Pascal (Clermont 2)	Master L.E.A.: "Production de Documentation Normalisée" (given over 1 year) Translation courses: N/A Writing courses: 140 hours Content (English only): Technical description, technical instructions, warnings, user manuals, online instructions, Web writing, visual communication, controlled English	Licence L.E.A.: Langues Etrangères Appliquées (given over 4 years) Translation courses: 192 hours Writing courses: 128 hours Content: (44 hours of Techniques d'expression); 86 hours in English: reports, summaries, syntheses, process descriptions, technical instructions and descriptions
Université de Paris 7 (Jussieu)	Master L.E.A. (over 2 years) (1) "Industrie des Langues et Traduction Spécialisée" Translation courses: 380 hours Writing courses: N/A Content: N/A (2) "Conception de Documentation Multilingue et Multimédia" Translation courses: 122 hours Writing courses: 170 hours Content: Principles of writing, benefits writing, technical writing	Licence L.E.A.: "Industrie des Langues et Traduction Spécialisée" (over 3 years) Translation courses: 200 hours Writing courses: 60 hours Content: Report writing, synthesis, abstracts, summaries
Université de Rennes 2	Master L.E.A.: "Traduction et Communication Multilingue" (over 1 year) Translation courses: 245 hours Writing courses: 80 hours Content: User manuals, online help guides	Licence L.E.A.: "Traduction et Communication Multilingue" (over 3 years) Translation courses: 220 hours Writing courses: 80 hours Content: Français Rédaction générale (SS), Français Redaction et relucture (S6)

Table 1. (Cont'd.)

Name of university	Master's level	Bachelors ('License') level
Université de Paris 12		Licence L.E.A.: "Ecrits spécialisés" (open to third-year bachelors students only) Translation courses: N/A Writing courses: 88 hours Content: Reports, synthesis, technical and professional genres
Université du Littoral Côte d'Opale	Master L.E.A.: "Langues et Technologies" (over 1 year) Translation courses: 220 hours Writing courses: 36 hours Content: Unknown	Licence L.E.A.: Langues Etrangères Appliquées (over 3 years) Translation courses: 288 hours Writing courses: 50 hours Content: Unknown
Université de Lille	Master L.E.A.: "Traduction Spécialisées Multilingue" (1 year) Translation courses: 200 hours Writing courses: 10 hours Content: Unknown	Licence L.E.A.: Langues Etrangères Appliquées (over 3 years) Translation courses: 140 hours Writing courses: 50 hours Content: Report writing (French, English, and 1 other language) and "Techniques d'expression"
Université de Bretagne Sud	Master L.E.A.: "Documents et Technologies de l'Information et de la Communication *No information available on the university Web site*	Licence L.E.A.: "Traduction – Rédaction" (over 3 years) Translation courses: 160 hours Writing courses: 40 hours Content: Unknown (3rd year only)
Université Jean Monnet (Saint-Etienne)	Master L.E.A.: "Traduction de Produits de Communication Multi-Supports" (over 1 year) Translation courses: 260 hours Writing courses: 10 hours Content: Unknown	Licence L.E.A.: Langues EtrangLres Appliquées au Commerce International (over 4 years) *No information available on the university Web site*
Université Technologie Compiègne (UTC)	DICIT: Concepteur – rédacteur de la documentation technique *No information available on the university Web site*	

The emphasis on writing instruction in these two programs, however, does not carry over equally at the master's level. While the master's program at Université Blaise Pascal currently offers 140 hours of writing instruction over one year (6.4 hours per week), the Université de Bretagne Occidentale offers only 30 (1.4 hours per week). Also, while one of the master's offered at Université de Paris 7 (CDMM) offers 170 hours of writing instruction, once again, this teaching is distributed over a 2-year period; the actual annual teaching load ends up to be around 85 hours (3.8 hours per week). Rennes 2 shows similar results, with 80 hours at the master's level (3.6 hours per week). A preliminary observation to be drawn from the data in Table 1 is therefore that written communication is typically not considered a central part of the training for future technical writers in France.

To a certain extent, the preference for translation training reflects the status of the L.E.A. programs in a country whose geopolitical context requires translation both into and out of French. Indeed, L.E.A. programs have long been at the forefront of translation training in France. Historically called a "bilingual bachelor's degree" (*licence bilingue*, Gallet-Blanchard & Peyronel, 2006), the purpose of an L.E.A. program is to prepare students to work in companies by giving them a solid foundation in intercultural considerations and at least two foreign languages (English plus one other language), while teaching them about the business environment they will work in (law, economics, management, marketing, etc.).

In this regard, the traditional emphasis on translation in an L.E.A. program creates a situation in which training for future technical communicators is already inherently intercultural and international in scope. Despite the potential advantage this training may offer future professionals for integrating an increasingly globalized work environment, however, French students tend to be weaker in writing tasks and the pragmatics of writing than they should, simply because they have had less experience in recreating the real language tasks that would prepare them for communicating in the workplace.[5]

[5] This comment is based on my own observations as a writing instructor in English. In effect, in the 6 years I have taught a third-year writing-intensive Business English course in our L.E.A. Bachelors program at Université Blaise Pascal (Clermont 2), I have observed that while in theory, our students have had prior experience (in French) writing various professional genres during their first 2 years (*techniques d'expression*), this experience proves to be ineffective because in the writing-intensive Business English courses. In effect, most students have sizeable difficulty in adequately adapting their genres to the different rhetorical situations they encounter in task-based scenarios (Dressen-Hammouda, 2004, 2006, to appear). Their difficulty in carrying out such tasks in English reveals their inexperience in writing audience-centered texts in their first language. Moreover, although it is always somewhat tenuous to make generalizations based on limited personal observations, I do not believe it unreasonable to expect that the situation is any different for students coming out of other L.E.A./technical writing programs in France, given that even less writing instruction is usually provided at most other institutions (see Table 1).

IMPETUS FOR INTERCULTURAL TECHNICAL COMMUNICATION IN FRANCE

Despite these challenges, this is an exciting time for intercultural technical communication educators in France. Most importantly, the strategic importance of having well-trained technical communicators is increasingly recognized by both multinationals and small and medium-sized enterprises (SMEs) throughout France and the E.U. This, in turn, provides impetus within European and national institutions to support writing instruction and programs in technical communication. For example, a recent survey of SMEs across Europe (ELAN, 2006), commissioned by the Directorate General for Education and Culture of the European Commission,[6] found that over the preceding 10-year period, European SMEs had undergone a significant loss in business as a result of workers' poor intercultural communication skills. Based on the respondents' answers, the study estimated that at least 945,000 European SMEs had been losing trade for this reason. Because strong intercultural and language training was viewed by the study authors as key for stimulating the European economy, they recommended that the educational systems and governments of member states invest more to develop adequate, government-led language policies, namely through university-based exchange, or Erasmus, programs (p. 40).

Of course, a greater investment in language development and the fostering of intercultural awareness are reflected in the various reforms recently implemented across the university systems of the E.U. In June 1999, for example, 29 European ministers signed the Bologna Declaration,[7] which stated that in order for higher education in Europe to prepare European university students to be interculturally sensitive multilingual citizens, better able to compete internationally, member states' higher education systems needed to facilitate university exchanges (Erasmus and Socrates programs) by realigning themselves upon a single model: LMD. The so called LMD reform sought to harmonize university structures across the E.U. by imposing a 3-year bachelors degree (L), a 2-year master's degree (M) and a 3-year doctoral degree (D). Such reforms are key for the development of intercultural technical communication education in France because they encourage L.E.A. programs to maintain high standards in

[6] The report, entitled "Effects on the European Economy of Shortages of Foreign Language Skills in Enterprise," was prepared by the UK National Centre for Languages (CILT) in collaboration with InterAct International and an international team of researchers. Its objective was to provide the Commission and decision makers in European member states with practical information and analysis of the use of language skills by SMEs and the impact on business performance.

[7] http://www.bologna-berlin2003.de/pdf/bologna_declaration.pdf
www.bologna-berlin2003.de/pdf/bologna_declaration.pdf

their existing intercultural and language training, while providing increased curricular opportunities for writing support.

In effect, the LMD reform has provided important opportunities for writing support at the university level in France. Within the new curricular structure resulting from the reform, the French government has increased its funding of English-language and writing instruction to first-year students (EEO modules, or *expression écrite et orale*). It has further proposed funding for projects targeting intervention and support for first-year university students, including writing support in general. As a reflection of this change, a government-funded project is currently underway to study university student writing at three universities in four disciplines (C. Donahue, personal communication, November 2006). The project builds on an extensive body of literature about the specific obstacles students at French universities face with reading and using sources, placing themselves in the academic conversation, learning new genres, and other relevant writing issues. Yet another university (Université de Paris 3) has begun to test all incoming students and offer writing courses to those who need them.[8]

The LMD reform has also created a window of opportunity for increased writing support in the new intercultural technical communication program to be presented in the following section. As a result of the reform, which extends master's-level study from 1 year to 2 years, French programs in technical communication will have more room to develop writing-intensive modules. As of the fall of 2008, however, only two master's programs intend to offer comprehensive, 2-year training programs in intercultural technical communication, including intensive writing instruction (Université Paris 7 and Université Blaise Pascal). In essence then, the challenge for intercultural technical communication training in France is not to make programs in technical communication more intercultural, for this quality has long existed due to France's geopolitical context and the influence this has had on national education policies (e.g., L.E.A. programs). Rather, the challenge today is to sufficiently prepare French students for a career in which written communication skills are central in French, English, and other languages.

The teaching of technical communication in France thus finds itself at a crossroads, although for different reasons than in North America. Clearly, being a well-trained technical communicator in France today implies having extensive knowledge about intercultural communication. Given France's current geopolitical and commercial context within Europe, technical communicators working in France for French companies or multinationals must necessarily possess the ability to localize and translate information effectively to a variety of audiences of differing national origins. At the same time, however, given the

[8] For more information, see http://www.prelude.in2p3.fr www.prelude.in2p3.fr. This program is further discussed by Donahue (2004).

long-term absence of explicit writing instruction in postsecondary education in France, special emphasis still needs to be placed on teaching students the value—and to some extent, the basics—of written communication skills in English, French, and other languages.

A TECHNICAL COMMUNICATION MASTERS PROGRAM AT UNIVERSITÉ BLAISE PASCAL

The 2-year master's degree in intercultural technical communication described in the following section has replaced a former 1-year program already in place at Université Blaise Pascal, called *Production de Documentation Normalisée* (see Table 1). While the former program has shown itself to be successful in many respects, the new master's program, called *Langues et Communication Technique* (MLCT), intends to broaden its recruitment base by strongly reinforcing instruction in written communication, intercultural communication, foreign languages, and the business environment.

Based on the preceding discussion about the French university context, the following specific needs have been identified for technical communication education in France and, as a result, for MLCT's curriculum:

- Providing vast practice in writing different genres for different audiences
- Developing rhetorical awareness in using different genres (audience, situation, purpose, genre, style)
- Developing skills in user-centered communication
- Developing a professional identity through writing
- Developing and reinforcing an understanding of genres as dynamic discoursal forms
- Continuing to develop an advanced awareness of intercultural differences, practices, and communication
- Combining specialized discoursal knowledge with IT tools
- Grounding discoursal and technical knowledge within the business environment

Such curricular needs are intended to promote the growth of intercultural technical communication education in France by both continuing to build on students' existing intercultural awareness and competence in foreign languages and by providing intensive training in the various aspects of written communication (roughly 240 hours), IT tools, and the business environment.

The master's program is designed to train students to work as specialists in intercultural technical communication by giving them expertise in company operations, knowledge management, IT tools, specialized foreign languages, and international business. These objectives are covered over a 2-year program, with approximately 400 annual teaching hours for each year of the master's.

Each year focuses on specific aspects of technical communication and allows students to apply their coursework during two specialized work placements in a French or international company.

Year 1

Introduction to technical communication	Procedural writing (descriptions, instructions, procedures, warnings), task analysis, user-centered texts, industry standards
Intercultural skills	Intercultural business and technical communication
Workplace communication	Report and proposal writing
English	Oral expression, English-French translation
Applied translation	German, Arabic, Chinese, Spanish, Italian, Portuguese, or Russian
Communication technologies	HTML programming Web site design Information architecture
Business environment	Project management Fundamentals of management Marketing European law Business strategies
Professional practice	Supervised work placement in international or French company (16 weeks minimum)

Year 2

Advanced technical communication	Online procedural writing, benefits writing, Web design and Web publishing, online help, training simulation videos
Languages & Communication	Controlled languages Applied translation Visual communication & visual design Oral communication

Communication theories	Accessibility and human factors User-based design Usability studies and user interview techniques Promotional language and advertising
Business environment	International project management Company knowledge management Competitive benchmarking Quality assurance Localization
Communication technologies	Collaborative authoring tools Metadata and database management (Structured XML, Content management systems) Computer-assisted translation Translation management systems Desktop publication software
Professional practice	Supervised work placement in international or French company (16 weeks minimum) Design project

Redesigning an L.E.A. program's traditional emphasis on translation to have it focus more closely on writing, while retaining its emphasis on intercultural differences and the business environment, is an interesting move for at least three reasons:

1. Being housed in a business-oriented faculty allows French students to continue acquiring the essentials of the business environment. Students trained as business-savvy communicators, involved with process design and usability testing, trained in the latest IT trends, will be valuable assets to companies, allowing companies to save both time and money;
2. A heavy emphasis on writing instruction, notably through a rhetorical and action-based approach to technical communication, will allow French students to refine and develop their "dormant" writing skills.
3. The reinforcement of writing instruction will allow for students to become accomplished technical writers, in addition to being translators. This choice reflects recent trends on the job market in France, where calls are increasingly made for technical writers capable of producing a variety of technical and professional documents.

WORKING WITH FRENCH TECHNICAL COMMUNICATORS

As a new generation of technical communicators are trained in France, they will increasingly come into contact with technical communicators from North America and elsewhere. In order to work and interact with them more effectively, a number of intercultural considerations must be taken into account. The technical communication professionals working in France and Europe who were interviewed for this chapter point overwhelmingly to similar considerations. First, the basics of the technical communicator's job are the same, irrespective of country or language. To work in an intercultural environment, it is important to know how to look, listen, and behave in different environments.

These skills can be taken a basis for the following behaviors, which are crucial to working more effectively in and with France's intercultural business environment.

1. Be polyvalent; be willing to multitask:
   Writers in France and Europe often find themselves doing a great deal of multitasking in their day-to-day work. Due to the globalized environment in which their companies typically operate and the accompanying localization concerns, French technical writers are often involved in document/quality control and budgets, in addition to writing technical documents.
2. Be aware of localization needs:
   Due to the geopolitical and cross-linguistic contexts in which they work, technical writers in France and Europe must also constantly adapt their communication and styles to various local markets. In this sense, it is important to identify what is, for example, specific to North American markets, what is international, and what is specific to other regional markets. Such local differences go beyond metric or imperial measurements and paper size. They include:
   - **Worldview**: One of the biggest problems cited by technical communicators in working with some U.S. writers or editors is that it is often assumed that what is done in the U.S. operates in the same way in the rest of the world. Customers from France and Europe do not necessarily want the same things as North American customers.
   - **ISO standards**: Europe overwhelmingly uses ISO standards; French and European writers still encounter difficulty in having EU directives (i.e., European law) implemented in documents produced in the United States.
   - **Legal issues**: Legal issues tend to be different in France and the United States.
   - **Language issues**: The problem of language goes well beyond obvious translation problems by also including how one distributes multiple language versions of documents in a global market. This becomes difficult

because, for example, the process of document distribution differs from one language to another.

3. Learn to speak the local language:
   Overwhelmingly, all the technical communication professionals consulted during the survey stress the same point: to work more easily with French technical communicators, developers and engineers, it is crucial to learn how to speak French. As a general rule, people tend to cooperate better when things have been made as easy as possible for them; here by speaking French.
   - When the French see you are making an effort at communicating with them in their own language, they become more open and willing to cooperate. It is a well-appreciated courtesy, even if your French is basic.
   - Not only is it polite, it often easier to obtain information from French speakers in French. To be more effective, interviews and investigations should be conducted in French.
   - Even if the French speakers who work in technical communication in France often speak good, if not excellent, English, most still prefer to communicate in French.
4. Maintain a respectful attitude:
   When North American technical communicators are dealing with their French colleagues, another extremely important consideration to keep in mind is to remember to treat them as equals.
   - Be patient and allow for extra time. If there are questions about the services or products being documented, it may not be possible for the French technical writer to answer them immediately. If the technical writer needs to go back and interview the source, it may take some time. In France, carrying out a task that relies on the input of several people tends to take longer than in the United States.
   - Avoid assuming that because someone has English as a second language that their written English will be poor. Many French technical writers do appreciate having their written English edited, but only when it is done so respectfully.
5. Be aware of behavioral differences:
   It is also important to observe French colleagues to see what forms of behavior are acceptable for different situations.
   - Personal space in France, as well as in other parts of Europe, is smaller than it is in the United States. People tend to stand closer together, such as when they are waiting in line (making them appear pushy) or when talking to one another.
   - People in France also tend to touch one another more. There is more kissing and hand shaking as a greeting, especially the first time colleagues see one another in the morning.

CONCLUSION

Today, technical communication is still a new field in France. Apart from the programs in technical communication offered at the Université de Paris 7, Université de Rennes 2, and Université Blaise Pascal, however, training for French technical writers is still largely translation-based. Nonetheless, there is now a strongly recognized need among French companies for well-trained technical communicators, able to cope not only with the usual translation tasks, but also with user-based communication tasks, accessibility, usability studies, localization, local and international project management, knowledge management, and IT technology. According to one Paris-based business, which develops and supplies open-source software to the largest French companies and administrations, the current need for trained technical communicators in France is such that without them, French companies will soon be forced to import technical communicators from other countries. To meet this urgent need, French universities must quickly adapt by offering communication-rich programs in technical communication, rather than translation-training only, in order to provide the wide range of skills needed in the profession today.

Clearly, this is a time of opportunity and change, crucial for the promotion and development of technical communication in France. In the coming years, ongoing cooperation between various university programs, namely between Rennes 2, Paris 7, and Blaise Pascal universities, can only reinforce and vitalize the training and professionalization of students in technical communication in France.

REFERENCES

Adam, J. M. (1992). *Les textes: Types et prototypes*. Paris: Nathan.

Altet, M. (1997). *Les pédagogies de l'apprentissage*. Paris: P.U.F.

Astolfi, J. P. (1993). *L'école pour apprendre*. Paris: E.S.F.

Bautier, E. (1995). *Pratiques langagières, pratiques sociales: De la sociolinguistique à la sociologie du langage*. Paris: L'Harmattan.

Beaufort, A. (1999). *Writing in the real world: Making the transition from school to work*. New York: Teachers College Press.

Burgel, G. (2006). *The university, a French dilemma*. Paris: Hachette.

Charlot, B., Bautier, E., & Rochex, J. Y. (1992). *Ecole et savoir dans les banlieues et ailleurs*. Paris: A. Colin.

Donahue, C. (1996). *French students' writing ability as a method of transcending the boundary into post-secondary studies*. Paper presented at the Conference on College Composition and Communication, Milwaukee, WI.

Donahue, C. (2000). *Genres, textual movement and subjectivity in learning academic writing: Interpreting the discourse of French and American student writers*. Unpublished PhD Dissertation, University of Paris 5.

Donahue, C. (2004). Writing and teaching the disciplines in France: Current conversations and connections. *Arts and Humanities in Higher Education, 3*, 59–79.

Dressen-Hammouda, D. (2004). Directed communicative modeling: Contributions of the genre approach to teaching professional English. *Les Cahiers de l'APLIUT, 23,* 89–102.

Dressen-Hammouda, D. (2006). A challenge for teaching LSP: Mitigating text and the need for multimodality. *Les Cahiers de l'APLIUT, 25,* 60–73.

Dressen-Hammouda, D. (2008). Aligning EAP writing pedagogies across European universities: A case study from France. In I. Fortanet-Gómez & C. Räisänen (Eds.), *Toward a harmonized European higher education: English as the medium of teaching and learning* (pp. 75–96). Amsterdam: John Benjamins, AILA Applied Linguistics Series.

ELAN. (2006, December). *Effects on the European economy of shortages of foreign language skills in enterprise.* Retrieved June 16, 2007, from http://www.cilt.org.uk/key/elan_finalreport.pdf

Fayol, M. (1995). Des modèles de production du langage à l'étude du fonctionnement du scripteur enfant et adulte. In J. Y. Boyer, J. P. Bionne, & P. Raymond (Eds.), *La production de textes, vers un modèle d'enseignement/apprentissage de l'écriture.* Montréal: Logiques.

Fayol, M. (1996). Apprendre à produire des textes. In C. Barré de Miniac (Ed.), *Vers une didactique de l'écriture.* Brussels, Belgium: De Boeck et Larcier et I.N.R.P.

Fayol, M. (1999). L'apprentissage de la lecture et de l'écriture. In J. A. Rondal & E. Esperet (Eds.), *Manuel de psychologie de l'enfant.* Brussels, Belgium: Mardaga.

Fayol, M. (2004). Les difficultés de l'orthographe. *Cerveau & Psycho, 3,* 2–5.

Gallet-Blanchard, L., & Peyronel, V. (2006, January 13–14). *La place de la recherche et de la reconnaissance institutionnelle en LEA/LANSAD.* Paper presented at Journées d'études sur le thème LEA/LANSAD: Convergences/ Divergences, Nantes, France.

Garcia-Debanc, C. (1993). Peut-on programmer les apprentissages en production d'écrits? In A. Bentolila (Ed.), *Les entretiens Nathan actes III: Paroles, écrit, image.* Paris: Nathan.

Haas, G., & Lorrot, D. (1987). Pédagogie de texte descriptif. *Pratiques, 55,* 5–23.

Lahire, B. (1993). *Cultures écrites et inégalités scolaires.* Lyon, France: P.U.L.

Miller, C. (1984). Genre as social action. *Quarterly Journal of Speech, 70,* 151–167. Second publication in Freedman, A., & Medway, P. (Eds.). (1994). *Genre and the new rhetoric.* Bristol, PA: Taylor & Francis.

Piolat, A. (Ed.). (2004). *Ecriture. Approches en sciences cognitives.* Aix-en-Provence, France: Presses Universitaires de Provence.

Piolat, A., & Pélissier, A. (Eds.). (1998). *La rédaction de textes. Approche cognitive.* Lausanne, Switzerland: Delachaux & Niestlé.

Plane, S. (2003). L'écriture et son apprentissage à l'école primaire. *Repères*, 26–27.

Reuter, Y. (1996). *Enseigner et apprendre à écrire.* Paris: E.S.F.

Thin, D. (1998). *Quartiers populaires: L'école et les familles.* Lyon, France: P.U.L.

CHAPTER 9

Teaching Technical Communication in India

Makarand (Mak) Pandit

The technical communication industry in India is growing at a fast pace. For example, when the Society for Technical Communication India Chapter (www.stc.org) was formed in 1999, it had a mere 15 members. Today, chapter membership totals more than 100 (Home, STC India, 2010). Moreover, the chapter's 2006 annual conference (held in Bangalore) attracted more than 500 participants from all over the country (Home, 8th STC India Annual Conference, 2006). These numbers indicate the growing demand for both technical communicators and related technical communication courses in India. This chapter provides readers with an overview of both technical communication practices and education in India in order to provide non-Indian practitioners with background knowledge that can help them work more effectively with Indian colleagues. It also provides technical communication educators with a foundation for teaching students about Indian technical communication practices, so those students can more effectively participate in a global workforce. However, before we look at this profession and its growth in India, let's quickly look at what triggered it.

THE ECONOMIC AND BUSINESS BOOM IN INDIA

The government of India adopted liberal financial policies and opened up its markets in the early 1990s. The winds of globalization carried the news about the Indian economy's potential to the entire world. Such news soon resulted in Indian companies marketing their services globally, and multinational companies began looking at India as an important emerging market for their goods.

The political equations changed quickly after the end of cold war, and more and more international companies became eager to cash in on the benefits of the Indian market. The successful nuclear tests carried out by India during those years, combined with the restraint India demonstrated in its military activities, convinced the world that the nation was now a matured democracy and a stable economy/place to do business. This image was further bolstered by then-President Bill Clinton's 2000 visit to India. The presidential trip sent a message that India was a safe country to invest in. The later start of the war on terror further strengthened U.S. ties as it began looking to India as a strategic partner and a strong global economic player.

While all this was happening, Indian engineers saw demand for their services increase, along with a boom in Indian high-tech industries. The number of Indian engineers and IT professionals employed abroad also increased manifold. Numerous international companies began looking at India as a major source of graduates trained in a variety of engineering and information technology (IT) areas. The U.S.'s Silicon Valley, moreover, was already full of well-settled Indians who were eager to take the plunge as entrepreneurs to launch several successful startups in India. These expatriates knew firsthand of the value of India's talent pool. Their business models, in turn, leveraged their base in India as a back office (which I prefer to call their "business engine").

The output of these "engines" in India was being delivered to the rest of the world through the strong channels of enterprises built by nonresident Indians living and working in other nations. These entrepreneurs could take this step because their risk was limited, and funds for such ventures were available. While all this happened, a number of larger Indian companies also made the most of the "talent advantage" being tapped by expatriate businesspersons. These Indian companies consolidated India as their base and delivered services to a wide range of clients around the world. The parallel developments in networking and online communication technologies made real-time business support for such enterprises possible from anywhere in the world.

The cost of setting up business and acquiring the talent through local or overseas recruitments was also increasing all over the world, making the relatively inexpensive costs of India-based expertise a key competitive advantage. Further contributing to this situation was the increasingly complicated international employment situation resulting from new visa and other employment restrictions imposed by nations such as the United States. As a result, most major multinational companies went in for the obvious solution that was within their reach—starting their own branch office in India.

The relatively lower cost of set-up, the availability of highly trained talent, and a stable democracy created the ideal combination that any business would want to have. The offshoring revolution had begun!

THE JOB BOOM

The pioneering work of Indian IT companies has led to two distinct advantages that affect technical communication in India. First and foremost, Indian talent got an opportunity to prove its capability in the IT sector. Second, more opportunities were created in other related fields, often termed as IT-enabled services, including business process outsourcing (BPO). These factors proved crucial in relation to slowing the "brain drain" India had been experiencing over last few decades. (A large number of well-educated Indians left the country in search of better opportunities.) Instead, the increase in the number of well-paying jobs in India and a simultaneous decrease in number of good jobs available in the countries where they resided has started a trend of reverse brain drain. In fact, a growing number of highly trained Indians are returning to India to take up the challenges and opportunities created by the economical growth. Many of them, in turn, are earning the same or comparable to what they were earning abroad.

Additionally, the exposure these individuals have had while working for big companies is helping them in their roles as managers or entrepreneurs in India. These factors are also making key contributions to India's work culture. Now, there are numerous and well-paying job options available to Indians, many of which were unheard of in the earlier generations, and some of them easier to adapt to, given the existing education system. But many of these positions are so different that there are no ready-made or compatible courses available to prepare prospective employees for them.

Technical communicator is one such job. At present, India needs to develop world-class courses meant to prepare the students to become the technical communication professionals so desperately needed by India's expanding high-tech sector.

SHIFTS IN INDIAN HIGHER EDUCATION

In the past 2 decades, Indian universities have created numerous courses and educational programs related to computers and the growth the IT industry has experienced. For example, today, Indians can get a Master's degree in Computers Management or a Master's degree in Computer Applications from certain Indian universities. Moreover, to address the growing needs of India's IT industry, all engineering courses taught at Indian universities now include computer use in their syllabi and are teaching Indian engineering students programming basics from the very first year of their studies. Additionally, there is a trend in which Indian teenagers are enrolling in a college and a computer institute at the same time. When these students complete their education, they will have simultaneously completed a computer course that can give them a better opportunity to land a job in IT. There are also a number of private institutes and public universities that have successfully launched special courses in networking, administration, programming, testing, and so on. The placement record of graduates from such programs reveals that these individuals are getting good jobs upon completion of their formal studies.

TECHNICAL COMMUNICATION IN INDIA

Technical communication in India is experiencing a similar kind of rapid job growth that is affecting higher education practices. To begin, the number of technical writers in India is increasing every day. This increase is due to a number of factors, including

- Increasing demand in local industry
- Increasing demand in outsourced jobs in IT, engineering, and other sectors
- Increase in knowledge process outsourcing (KPO)

In the 2005 salary survey done by the STC India Chapter, it is mentioned that only 15% of participants have had any formal education or training in technical writing (Paresh, 2005). This percentage went up to 18% in the 2007 salary survey (Paresh, Pandit, & Ramnathan, 2007). The author feels that with the recent start of private institutes and formal courses in technical communication, this overall percentage should have increased much more. It is possible that due to the large number of professionals entering the field, this percentage increase is only marginal. This effect is also likely because of the fact that the total number of participants of the survey decreased from 455 in 2005 to 281 in 2007 (Paresh, 2005; Paresh et al., 2007). One more point is worth noting here. As reported in the 2007 survey, candidates having formal training in technical communication are paid 12% more (Paresh et al., 2007). All of these factors have led to the rise of formal technical communication education in India. Perhaps a key example of this development is the Post Graduate Diploma in Technical Communication now available to India's current and prospective technical communicators.

THE STARTING OF PGDTC

A bit of history of how the Post Graduate Diploma in Technical Communication (PGDTC) started would be helpful in understanding its structure. After reading the information about STC India Chapter's 2003 Annual Conference, the University of Pune developed interest in offering formal training in technical communication. The author, a practicing technical communicator, was invited to discuss the possibility of starting a course in technical communication for students at the university. To achieve this objective, the author invited his co-professionals, Dr. Sunil Gokhale and Frederick Menezes (both working as technical communicators), to join the course-development team. Together, these three practitioners developed an outline for the very first technical communication course offered in India. It was a 1-month-long course that was structured as an extended workshop that covered important topics in technical communication, including technical writing and technical editing.

Encouraged by the response of university administrators, the three course co-creators then revised the overall content of the class to transform it from a 1-month- to a 2-month-long workshop. This transition included sessions that focused on hands-on training with documentation and layout tools and the use of an actual technical communication project from industry. After the second time the course was offered, the university was convinced that a full-fledged postgraduate diploma in technical communication would be viable. To work toward this goal, the university appointed a subcommittee to study technical communication programs offered by other universities in order to develop one that would address topics covered in such classes while also maintaining a job-oriented focus. The result of this examination was the Post Graduate Diploma in Technical Communication (PGDTC) currently offered by the University of Pune.

PGDTC COURSE CONTENTS

The resulting PGDTC Diploma program requires students to produce a total of seven papers, each of which focuses on different aspects of technical communication (see Table 1).

Table 1. Papers Included in the Post Graduate Diploma in Technical Communication Program

Paper No.	Name	Credits
CTC 01	Introduction to Technical Communication	2
CTC 02	Information Design & Development	2
CTC 03	Technical Writing	4
CTC 04	Grammar & Editing	2
CTC 05	Technical Communication Software Tools	2
CTC 06	Project Work	2
CTC 07	Advanced Technical Communication Topics	2
Total Credits		**16**

Each paper in turn had different educational objectives that, together, provided participants with the knowledge base and skillsets needed to succeed in the field of technical communication. The various papers of the PGDTC program and the related educational objectives of each course are described here.

CTC 01: Introduction to Technical Communication

This paper provides students with an introduction to technical communication through a survey of its theoretical foundations, history, core competencies, opportunities, and challenges. The paper project also includes an introduction to different kinds of technical communication documents or products.

In drafting this paper, students are required to focus on two main areas:

Area 1: Communication theory: communication models, frames of reference, forms and styles; textual communication versus visual communication; online communication versus print communication.

Area 2: Technical communication: communicating technology; need for technical communicators, roles and functions of technical communicators, skills required for technical communicators; history of technical communication, technical communication in India; challenges in technical communication; different kinds of technical documents (print and online).

CTC 02: Information Design & Development

The focus of this paper is the information development life cycle. The students are required to include a detailed look into each phase of the cycle. The paper must also include a study of various information organization structures, factors affecting information and document design, and strategies for information organization. Additionally, students explore differences in information design and writing for print and online media.

To achieve these objectives, students are asked to focus on six particular topic areas:

Area 1: Information development life cycle

Area 2: Planning: purpose and information requirement analysis, audience analysis, scope; information delivery medium

Area 3: Research: brainstorming, the 5 Ws and 1 H; interviewing (face-to-face, e-mail, telephone), note-taking, Internet research, library research, using questionnaires; documenting sources

Area 4: Information design: organization structures; outlining and chunking; document layout and design (print and online), cognitive, affective, and physical factors; typography; visuals

Area 5: Differences in information design and writing for print and online media

Area 6: Document prototyping and production

CTC 03: Technical Writing

This paper provides students with an in-depth look at the technical writing process, including forms of discourse, methods of development, writing and revising drafts, collaborative writing, and creating indexes. The paper requires students to include a detailed study of technical writing style and language. To achieve these objectives, students focus on four particular topics or areas:

Area 1: Writing process: rhetoric and composition; forms of discourse: description, exposition, narration, persuasion; methods of development: cause and effect, chronological, definition, division and classification, general and specific, inverted pyramid, order of importance, sequential, spatial; style and tone; sentence construction, paragraph construction; quoting and paraphrasing; writing introductions and conclusions, writing titles and headings

Area 2: Style and language: completeness and accuracy; unity and coherence: conciseness, pace, transition; sentence variety: emphasis, parallel structure, subordination; clarity: ambiguity, awkwardness, logic errors, positive writing, voice; word choice: abstract/concrete words, affectation and jargon, clichés and trite language, connotation/denotation, defining terms; style guides

Area 3: Collaborative writing: working in teams, handling conflicts, peer reviews

Area 4: Creating indexes: using a review of a text to create a related index

CTC 04: Grammar and Editing

The purpose of this paper is to provide students with an introduction to basics of grammar and a study of advanced grammar. In producing the paper, students are required to include an in-depth look at technical editing and editing strategies to achieve an appropriate technical writing style. The overall class is structured around three central areas:

Area 1: English grammar: adjectives, adverbs, conjunctions, interjections, nouns, prepositions, pronouns, verbs, functional shift; case, gender, mood, number, person, tense; subject-verb agreement; modifiers; pronoun reference; sentence faults

Area 2: Punctuation and mechanics: abbreviations, capitalization, contractions, dates, indentation, italics, numbers, proofreading, spelling, symbols

Area 3: Editing: levels of edits; working with editors; editing to achieve appropriate style

CTC 05: Technical Communication Software Tools

This paper provides students with the chance to study select technical communication software tools including Microsoft Word, Adobe FrameMaker, and Macromedia RoboHELP.

The areas of focus in examining such tools include

Area 1: Microsoft Word: documents, styles, images, cross-references, drawings, tables, contents, indexes, captions, revision tracking, autotext, macros, templates

Area 2: Adobe FrameMaker: documents, books, styles, images, cross-references, drawings, tables, contents, indexes, templates

Area 3: Macromedia RoboHELP: Creating online Help; topics, links, popups, hotspots, contents, indexes, special topics, DHTML effects, tables, build tags

Area 4: Others tools: image-capturing tools, PDF converters

CTC 06: Project Work

For this paper, students participate in a team project that involves creating a printed user's guide or online help for a product. This approach involves using teams composed of a predetermined number of members who then work together under the guidance of a project guide. The guide is one of the faculty members of the course. The guide helps them right from scoping the project, assigning responsibilities to getting a sign-off. Most of the projects are sponsored by the industry and the project deliverable is ultimately used by the sponsor. Projects help the students to practice what they have learned and also to understand the interdependencies on team members in the real-life environments. Successful projects often lead to student placements.

CTC 07: Advanced Technical Communication Topics

In this capstone paper, students are given a brief introduction to advanced technical communication and related topics that will affect them at some level or other in the workplace. Topics in this class include usability, human factors, managing technical communication projects, time estimation, single sourcing, and localization.

Through this seven-paper approach, the PGDTC provides students with a strong knowledge base and foundational set of skills they can use to move into technical communication positions in a variety of Indian-based industries.

Figure 1 shows the structure of the PGDTC program, and Figure 2 shows the percentage of time spent on various papers. The percentages are calculated based on the record of actual time spent on the topics.

The structure of the PGDTC course mandates a specified number of hours be spent on the topics. However it does not impose a time limit. The faculty is given the freedom to conduct additional sessions, if need be. So the percentages mentioned above are different than those expected. The author feels that this is a natural process of trial and error and helps to fine tune the course.

Within this framework, a good deal of time appears to be spent on the topic of technical writing, yet the course assigns that topic the same number of credit

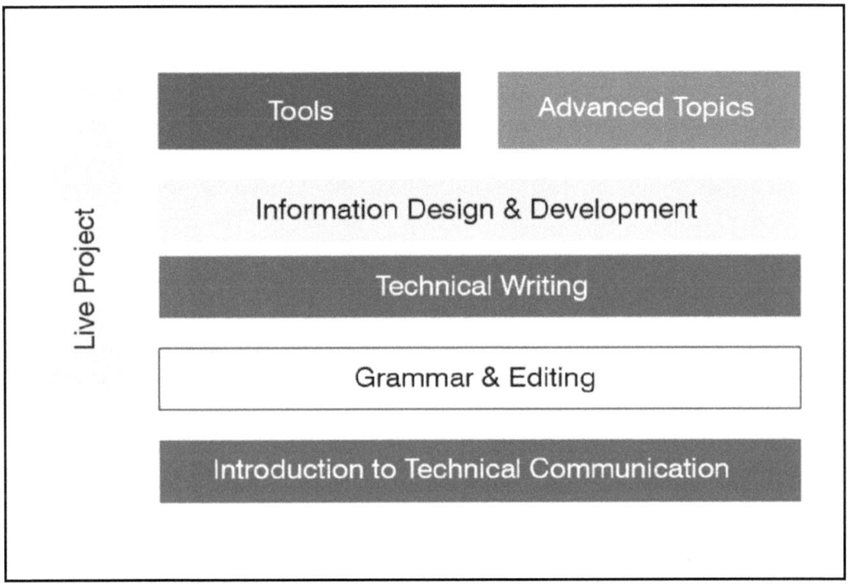

Figure 1. PGDTC course structure.

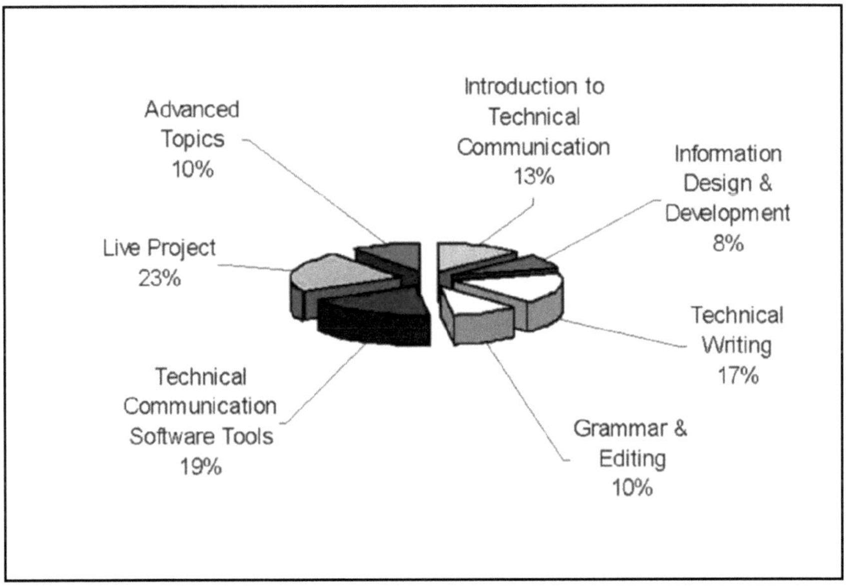

Figure 2. Overview of time spent on various subjects.

points as some of the other papers (for example information design and development). A lot of time in the course is also spent on learning tools (19%), and the maximum amount of time in the course (23%) is spent on the actual technical communication task, or the Live Project. The author feels that, within this framework, there is still some room to increase the number of hours spent on language, grammar, and editing. There should also be some session time spent on examining style guides and how to use them. (The topic of style guides was addressed in the course and included a discussion of the needs and uses of such guides, but the course did not cover the development of such style guides.)

The emphasis is not restricted to written communication only. Interpersonal communication and soft skills are also paid attention to. Students are encouraged to study different subjects and give presentations to their entire class. Though this is not a part of either the course or exam, the author feels it helps in personality and leadership development. Good presentation based on an extensive research gives the students the confidence that is sufficient to propel them in the initial stages of their career.

The course lays a lot of emphasis on assignments. The rating given to assignments is considered while giving term-work marks for all the subjects. The students are often asked to repeat the assignments so that they can improve their ratings, and most importantly their skills. The assignments are based on various phases of Document Development Life Cycle.

The PGDTC concludes with a written examination for the papers Introduction to Technical Communication, Information Design & Development, Technical Writing, and Grammar & Editing. There is a hands-on examination to rate the tools expertise acquired by the students, and for the project, an oral examination is conducted. Technical writing managers from the industry are invited to examine the projects (the actual deliverables of the group), talk to all the group members, and give grades in consultation with the respective project guides.

LANGUAGE AND COMMUNICATION

A variety of linguistic factors also affect technical communication education in India. For example, most Indians speak English as a second (or a third or fourth) language, and many Indians may have done their formal schooling in various Indian regional languages (like Marathi, Kannada, Gujarati, and so on). Thus, when such students write English, they may be assembling the sentence in their first language and then translating it into English to produce the first draft of a document. It is then natural for such students to review and revise this initial English draft later and make heavy changes. If they do not perform such revisions, then editors will need to do it for them. In this way, documents produced by Indian SMEs often require the editorial assistance of technical editors or technical writers to create effective final documents for a wider reading audience.

To provide technical students with the skills related to such a situation, the Grammar & Editing paper of the PGDTC program includes an introduction to the basics of grammar and a study of advanced grammar. The paper also includes an in-depth look at technical editing and editing strategies to achieve an appropriate technical writing style.

Considering many Indian technical communicators will actually be working for multinational corporations or within global contexts, there is a growing need to provide Indian technical communicators and technical communication students with training in intercultural communication. In the last few years, a number of companies have emerged to address this need, and professionals working in the IT and the BPO sectors have been pursuing such training for some time. The challenge now becomes integrating such intercultural communication training into a technical communication curriculum. For this reason, the PGDTC program might wish to consider giving intercultural communication a stronger or a more prominent focus in its 7th paper area of Advanced Topics in Technical Communication. (At present, that paper covers topics such as international English and localization, but balances the treatment of those topics with discussions of topics such as usability and single-sourcing.)

THE PATH FORWARD

Industry's response to Pune University's PGDTC program indicates the wide acceptance of the program and its prospects for success. Individuals who participate in the Pune PGDTC program have reported a roughly 65% job placement after completing the program. This percentage, moreover, is expected to increase with the recent launch of the new Post Graduate Diploma course. At the time of this writing, 97% of the students enrolled in this course had already been placed into industry positions because of the training acquired in the course (it is worth noting that many students were receiving job offers before actually completing the course).

These success rates, however, do not mean the PGDTC course is perfect. Rather, as with all newly developing programs, it needs to evolve further as instructors, students, and the technical communication field in India learn more. In fact, the university will need to keep the changing nature of Indian and global business scenarios in mind as the class develops and changes over time.

As per the 2007 STC India Chapter Salary Survey, 17% of participants have graduated in Arts with English Language as a major, whereas 59% participants came from an Engineering or Science background (Paresh et al., 2007). This statistic underlines the need for technical communication courses in India to have more sessions dedicated to language.

Intercultural training is also very important, as an increasing number of Indian companies are serving as back-end global offices for a growing number of multinational organizations. It is particularly important, because a wider range of nations and cultures from North America, Europe, Australia, and Asia are turning to India to meet a variety of business needs. At the same time, Indian companies in sectors such as steel, engineering, pharmaceutical, telecom, IT, and biotechnology are gaining in their global presence, and Indian management institutes, in turn, are increasingly working with multinational corporations that wish to use their campuses as a base for recruiting managers for global projects. From an intercultural perspective, these developments mean that Indian professionals are leading from the front (e.g., management), while simultaneously working from behind the stage (e.g., business process outsourcing). Indian educational institutes will need to include an increased global element in all their courses. As technical communication plays an important part in this new situation, the formal education will also need to address this growing global context through different classes and training.

Management training will be another important area. According to the 2007 STC India Salary Survey, 56% participants mentioned that their team size increased in the last year (Paresh et al., 2007). Almost 85% of respondents indicated that they worked as a part of technical writing teams (Paresh et al., 2007). With technical communication being a highly specialized field, most companies will prefer to promote experienced technical communicators to manager's roles rather than getting in professionally trained managers. In the same survey, 15% of participants indicated that they spent 50% or more time in managerial activities, and only 10% of the 100 participants had any formal training in management (Paresh et al., 2007). The author feels that as the team sizes increase, the demand for managers will also go up. So adding management training, more specifically, project management training, to the courses seems to be the next logical step. However, this training should not be restricted to conventional project management. It should take into account the latest effects of trends like single-sourcing, content management, and DITA on Technical Writing Project Management.

Today Indian talent is considered to be one of the most competitive in the world. But things will not stay like this forever. The 2007 Salary Survey shows a comparison of Indian and U.S. salaries (Paresh et al., 2007, p. 32). Based on the data available, the author has extrapolated trend curves, which indicate that with the current rate of increase, the Indian salaries will intersect the U.S. salaries in 2014. This intersection will most probably not happen, but it is certain that the salary curves will come fairly close. If Indian technical communicators want to remain competitive, they will have to learn and acquire new skills, and do that very fast.

The message is clear. If India wants to retain the edge, it will have to create better courses that can help its talent to be globally competitive.

REFERENCES

Course Syllabus, Post Graduate Diploma in Technical Communication, Department of Communication & Journalism, University of Pune.

Home. (2006). 8th STC India Annual Conference @ Bangalore. Retrieved January 2, 2010 from http://stc-india.org/conferences/2006/home.htm

Home. (2010). *STC India.* Retrieved January 2, 2010 from http://stc-india.org/index.htm

Paresh Naik. (2005). *STC India 2005 salary survey.* Society for Technical Communication India Chapter. Retrieved from http://www.stc-india.org/activities/surveys/STC_India_2005_Salary_Survey.pdf

Paresh, N., Makarand, P., & Varadrajan, R. (2007). *STC India 2007 salary survey.* Society for Technical Communication India Chapter. Retrieved from http://www.stc-india.org/activities/surveys/STC_India_2007_Salary_Survey_Results.pdf

SECTION III

Connecting Instruction to Professional Practices: Merging the Workplace with the Classroom

CHAPTER 10

Between the Lines: Reading Contextually in the International Internship

Lyn F. Gattis

The international internship in technical communication—a short-term period of employment outside a student's home country—can be an exceptional learning opportunity for graduate students or upper-division undergraduates. Cross-cultural work experiences can help communication students build a sound base for future collaborations with clients, subject matter experts, and subcontractors from other cultures. An international internship can give them more of a nuanced understanding of translation and localization issues and make them more productive members of global teams during their careers. It can even prepare students to develop materials for subcultures within the dominant home culture by increasing their sensitivity to the information needs of different audiences. In short, the international internship can provide a first important step toward the "intercultural communication competence" (Barna, 1998, p. 187) technical writers, editors, and designers need in a global economy.

However, the success of international internships depends largely on three related factors: (a) the students' ability to engage in praxis by balancing academic theory and real-world practices; (b) their capacity for coping with culture shock, or the disorienting effects of being immersed in an unfamiliar environment; and (c) the students' skill at reading contextually, so that they perceive and respond to cultural differences appropriately, both in and out of the office. The purpose of this chapter is to offer guidelines that will help interns succeed in these important areas.

I begin the chapter by examining praxis in light of scholarship on culture shock and on contextual reading. For interns abroad, the issues are intertwined.

Next, drawing from my own experiences as an American intern in a software firm in northern Europe, I relate the international internship to the three stages of new employment proposed by Anson and Forsberg (1990) and by Lutz (1989): (a) expectation or pre-arrival in the workplace, (b) disorientation or breaking-in, and (c) resolution or acceptance within the new culture. Each stage overlaps with the process of working through culture shock and also poses a different set of contextual reading challenges for interns. Finally, I suggest several ways academic programs can prepare prospective interns to be more effective when they enter the global workplace.

LITERATURE REVIEW

Whether at home or abroad, students in technical communication internships continually balance two points of view. As developing practitioners who are learning to provide real deliverables to real employers, the interns must apply academic knowledge to real rhetorical problems. At the same time, as novice researchers who aspire to deep understanding of rhetorical principles, the interns must view day-to-day workplace interactions through the lens of communication theory. Internships thus offer students a focused opportunity to experience a sense of praxis, which Sullivan and Porter (1993) have described as "a research perspective willing to critique both theory and practice by placing both in dialectical tension, which can then allow either to change" (p. 226). In any internship, students likely encounter for the first time the need for professional praxis—the expectation that they balance what they have learned from coursework with the exigencies of an actual workplace.

When the internships take place outside the students' home countries, students encounter the need for personal praxis as well. Arriving in the host country with theoretical notions of the new culture, they must immediately begin adjusting to unfamiliar realities as well as the sudden absence of people, attitudes, and practices they have always known. Inside and outside the workplace, life in a new culture is a continuous, often stressful process of comparing the lived experiences of each day against preconceived ideas brought from home. Kalervo Oberg, the anthropologist who coined the term *culture shock* several decades ago, likens the phenomenon to a disease "precipitated by the anxiety that results from losing all our familiar signs and symbols of social intercourse. These signs or cues include the thousand and one ways in which we orient ourselves to the situations of daily life," including customs, facial expressions, words, and other signals, "most of which we do not carry on the level of conscious awareness" (1960, p. 177). Such stresses have also been characterized as cultural fatigue or cross-cultural adjustment (Eschbach, Parker, & Stoeberl, 2001) and as a form of transition shock felt more intensely because of the alien environment (J. Bennett, 1998). Although psychological research on culture shock has tended to focus less on short-term cross-cultural experiences and more on the long-term adjustments of entire groups, "common sense" and studies of exchange students "suggest that the greater the dissimilarity between two cultures, the more pronounced" the culture shock (Mumford, 2000, pp. 74–75). Like ethnographic researchers with an etic methodology, international interns always stand a bit outside the second culture (Geertz, 1983). However, like researchers with an emic approach, interns must also adjust well enough to meet certain insider standards as persons who live, work, and participate in the host culture. The unique nature of international internships thus requires the students to develop a type of praxis for daily life: balance between theory and practice that parallels and complements the praxis of the technical communication workplace.

Unfortunately, for purposes of dealing with culture shock and developing praxis, the advice commonly given to students preparing for study or internships abroad—to gather as much data as possible about the second culture—is, by itself, inadequate. Advance information does appear to reduce the number of cultural surprises newcomers encounter, thereby shortening the adjustment period (Eschbach et al., 2001). Such information can help interns to articulate research questions, acquire metalanguage for cultural analysis and description, and become aware of things to look for in both workplace and culture (Jorgenson, 1999). For example, matrices developed by Hofstede (1991) and Trompenaars (1993) can encourage students to think about the range of values, attitudes, and behaviors they may observe for the first time in a new culture. However, a preoccupation with cultural data can also create terministic screens that constrain interns in what they are prepared to see, fostering essentialist views of their host cultures. For example, the host company for my own internship worked with clients in several countries and conducted business in three languages. Over time I realized that my co-workers actually reflected traits of the global information technology community as much or more than the national characteristics I had expected to find. "Knowing exactly 'what to expect' often blinds the observer to all but that which confirms his or her image. Any contradictory evidence that does filter through the screens of preconception is likely to be treated as an exception and thus discounted" (Barna, 1998, pp. 178–179). An approach that overemphasizes "definitive culture" can easily become selective, failing to allow for the influence of economic, political, and historical factors outside the culture; and even a "culture-free" model of communication (which is difficult to achieve) can result in ethnocentric attitudes (Hunsinger, 2006, pp. 31–32). When interns find that people in country A or B are quite different from the students' preconceived profiles, the disconnect may negate whatever "anticipatory adjustment" (Eschbach et al., 2001, p. 272) students have tried to make, thus contributing to culture shock.

A complicating factor is that many Americans find it genuinely difficult to imagine that people in other cultures are all that different from themselves, especially since changes in technology and commerce seem to be making cultures more alike in some ways (Campbell, 1998). From the melting-pot perspective, dissimilarities are assumed to be superficial, while basic human nature is constant (M. Bennett, 1998). Consequently, Americans may believe that the most appropriate cross-cultural communication strategy is sympathy—remembering or imagining "how we ourselves might think or feel in similar circumstances" (p. 197), which can be misleading when the reality is difference rather than similarity. One example of sympathetic thinking is the minimal training many employees receive for cross-cultural assignments, a policy that reflects employers' assumptions of cultural universalism, even when many expatriates report needing more, not less, training (Suutari & Brewster, 1998). In addition to a mistaken belief in cultural similarity, Barna (1998) notes other related obstacles to genuine intercultural communication, such as assuming that words and phrases in another language have only one meaning; relying on cultural preconceptions and stereotypes; and evaluating the host culture hastily, using one's own culture as the norm. Interns abroad who assume cultural similarity find professional and personal praxis more elusive because they are unable to draw accurate inferences from the real-world practices they see around them.

By contrast, when interns assume differences among cultures, they are actually better prepared to consider situations from the point of view of their hosts. The relevant communication strategy becomes empathy, or "a shift in perspective away from our own to an acknowledgment of the other person's different experience . . . given different constructions and circumstances" (M. Bennett, 1998, pp. 208–209). Developing empathy does not mean relinquishing one's own values; rather, it means suspending the self long enough to allow one's imagination to understand the other person's purpose (M. Bennett, 1998). Consequently, instead of collecting factoids about the culture, interns prepare more effectively by first developing "an investigative, nonjudgmental attitude and a high tolerance for ambiguity—all of which require lowered defenses"; then the interns can consult the host country's history, literature, and art to develop a framework for what they will observe (Barna, 1998, p. 179). As an alternative to cultural data collection, Hunsinger (2006) offers the cultural theory of anthropologist and sociologist Arjun Appadurai, who identifies "people, media, technologies, capital, and political ideologies" as the critical influences on cultural identity (Hunsinger adds the influence of sacred values as well) (p. 40). To avoid oversimplifying another culture, Scollon and Scollon (2001) also emphasize the importance of keeping in mind both similarities and differences between groups, that is, basing cross-cultural comparisons "upon more than a single dimension of contrast" and remembering that no one individual in another culture "embodies all of his or her group's characteristics" (pp. 170–171). Barna (1998) stresses that culture shock "is the least troublesome to those who learn to accept cultural diversity with interest instead of anxiety and manage normal stress reactions by practicing positive coping mechanisms" (p. 186). Interns will be more flexible and better able to adjust to new surroundings and situations if they are prepared to find cultural differences.

If social and cultural data collection alone is inadequate preparation for working abroad, then how can prospective international interns best get ready? Simply put, the most helpful strategy for interns is to learn how to read: workplace texts, certainly, but mainly con-texts—audiences and situations in unfamiliar settings. Contextual reading skills are necessary on both professional and personal levels. Professionally, Anson and Forsberg (1990) argue that before any writer can fully understand an audience, take on a persona appropriate to the workplace, or produce useful texts, "the writer must first become a 'reader,'" through a dynamic process of analyzing, interacting, and adapting (pp. 207–208). For example, an experienced writer can "read" the situation in which technical procedures are likely to be problematic for a given audience. The writer will understand which comparisons will make the most sense, how concepts should be organized for those readers, and which graphics will aid comprehension. The contextual reading skill of experienced professionals is analogous to what Schon (1987) describes as "artistry," the ability to function competently in those "indeterminate zones of practice" where professional problems are less clear-cut than theory might indicate (p. 13). Schon suggests that we become artists as we learn to name problems and frame them contextually, so that (with our help) readers can draw accurate conclusions from the material we have written.

The principle of naming and framing problems contextually applies equally to interns in a cross-cultural workplace. In the beginning, students are likely to view the internship as a vehicle for knowledge transfer and honing of skills. They may imagine themselves learning by example, as they observe the "knowing-in-action" that professionals demonstrate almost instinctively as they solve problems (Schon, 1987, p. 33). The students may also believe that most of their assignments as interns will involve applying facts and rules from coursework when problems are familiar, and drawing from theory to devise solutions when problems are new. They may feel (not incorrectly) that they will be expected to demonstrate mastery of basic communication skills and possibly to bring additional technical content, metalanguage, and problem-solving expertise to the organization. New interns may believe that workplace tasks are well-formed, logical extensions of classroom experiences, with minimal ambiguity.

However, with appropriate mentoring before the internship begins, students can become aware of the important role of contextual reading and praxis in the workplace. Instead of preparing only for clearly defined, rule-based tasks, interns can learn to think critically about assumptions and strategies while outcomes can still be adjusted. Students can thus learn to distinguish between two types of problems: those that require simple application of principles, and those that call for students to respond by rethinking "strategies of action, theories of phenomena, or ways of framing the problem," followed by tests of the new approach's effectiveness in the moment (Schon, 1987, p. 35). In technical communication, for example, students who have learned to read contextually can evaluate documents for more than mechanics and grammar. They can see inconsistencies and redundancies that make reading more difficult. Instead of becoming preoccupied by microissues such as typography or margin widths, they can use design principles to make data easier to compare, thereby making thinking easier for the client (Tufte, 1983). They can step back from the text to consider questions a reader new to the topic might ask. In short, as they become more skilled at reading contextually, students learn to go beyond rules. They prepare themselves to work with the absence of a single correct answer, or with the possibility of several answers. As their contextual reading ability improves, they are better able to exercise *phronesis*, or good judgment about timing, rhetorical needs, and the relationship of *episteme* to *techne,* theory to application. The ability to engage in praxis by reading contextually could be categorized among the types of knowledge described by English and Jereb (1995) as "exciting," that is, skills well above the basics, exceeding employers' requirements (p. 66). Interns who read contextually keep in mind the audience's purpose for consulting the document in the first place.

Similarly, contextual reading skills also enable interns in a new culture to expect and cope effectively with differences on a personal level. Through the process of learning to name the phenomena around them and frame them contextually, students develop the ability to balance the general with the specific. They become more adept at deciphering linguistic, social, and environmental cues. For example, once students realize that a slight bend in a boulevard may signal a complete change of street name, navigation as a whole begins to make more sense. As they become aware of local norms of courtesy, small talk, and personal space, they become more sensitive to how their behavior may appear to others. Instead of finding fault with a culture different from that of North America, students become interested in how and why cultures develop in certain ways. As they learn to see themselves within the host culture rather than apart from it, they are able to balance the life they expected with life as it is, a type of personal praxis that complements and supports praxis in the workplace.

In the following sections, I discuss in greater detail the three stages of new employment, the aspects of each stage that relate particularly to contextual reading, and the steps in each stage that help to build professional and personal praxis. Table 1 provides an overview.

STAGE ONE: EXPECTATION

Ideally, prospective interns allow at least two full semesters for the *expectation or pre-arrival* stage, a period of getting ready mentally and psychologically. The long lead time reflects simple logistics as well. Negotiations with employers may take several months. Students also obtain passports, work out housing arrangements, and possibly learn some of the language(s) spoken in their host countries. In addition, students need time to practice visualizing themselves in

Table 1. Summary of Stages during an International Internship

Stage of internship	Contextual reading needs	Steps toward professional praxis	Steps toward personal praxis
Expectation (or pre-arrival)	Assess strengths, weaknesses, knowledge gaps, coping skills	Research employers carefully; negotiate employment terms	Practice exposure to unfamiliar situations; get comfortable with difference
Disorientation (or breaking in)	Examine assumptions about the host culture	Listen actively; learn through social interaction	Keep positive attitude toward new experiences
Resolution (or acceptance)	Pay attention to the nuances of the host culture	Respond appropriately to audiences; share in employer's success	Empathize; understand and display culturally appropriate behavior

new surroundings. To gain experience in naming and framing problems contextually, they make opportunities to encounter new and unfamiliar people, foods, situations, locales, modes of wayfinding, languages, and entertainment. They start learning to be flexible in the presence of difference.

In the expectation stage of the internship, students prepare effectively by assessing themselves honestly, practicing their contextual reading skills, and identifying and communicating with potential employers.

Taking Stock

Before investigating employment possibilities, prospective interns clarify for themselves what they hope to accomplish with an internship abroad. They evaluate their personal strengths and weaknesses, including such personality characteristics as their sense of security in their own identities, their ability to approach situations without judging, their skill at integrating seemingly unconnected bits of information, and their willingness to understand the perspective of another person (J. Bennett, 1998). They take stock of the inner resources they have developed in periods of adaptation or stress, and they also consciously test their coping skills in new situations. For example, they might join a new campus organization or volunteer for an unfamiliar function in a community group, reflecting on what they learn and how they feel in the process. They work on becoming comfortable with different types of people. If they have studied a language other than English, they visualize themselves immersed in a culture for which the other language is primary. They also look for opportunities to interact with individuals from other cultures, such as attending campus events with international foods and music, especially events that deemphasize American culture and spoken English. They also "practice body awareness" so that they understand when they are experiencing stress and what they can do about it (Barna, 1998, p. 187). In the expectation stage, students work on making sense of unfamiliar situations and honestly assessing their reactions.

Building Contextual Reading Skills

In this stage, students also consider carefully, perhaps with help from faculty or employers, their current abilities to read and interpret organizational needs in other cultures. Research now suggests a link between rhetorical effectiveness and the writer's understanding of the social environment (MacKinnon, 1993), including the reader's cultural framework (Campbell, 1998). Consequently, while generic writing skills are necessary and transferable from the classroom to the workplace, they are ultimately less important than the ability to cope with "the unfamiliar epistemological, social, and organizational characteristics of a new context" and to adapt content for different reading communities (Anson & Forsberg, 1990, p. 201). Similarly, while specific technical knowledge does matter in the workplace, "highly situational knowledge" is a necessary complement (p. 225). Therefore, during the expectation stage, an essential task for interns is to evaluate the strengths of their current situational reading skills.

The context for these skills need not be academic. For example, students consider occasions when they were able to "read between the lines" in previous jobs. They analyze specific cues that helped them bring about successful outcomes in a group project, a campus or community organization, or other social settings. They remember situations in which they had to resolve conflicts with peers or supervisors, or collaborate with individuals of different personality types to accomplish tasks. They reflect on their ability to interpret background factors in national or world events. Feedback from advisors or employers can help students to improve their abilities to read situations before they begin their internships.

Identifying Employers

During the expectation stage, students gain further experience in contextual reading through the act of negotiating the internship in the first place. Although interest in short-term employment abroad is growing in U.S. technical communication programs, established cross-cultural partnerships are still somewhat rare, and opportunities for internships may not be immediately obvious. Faculty and students must therefore read professional situations creatively in order to find job placement in other countries. The following resources are worth exploring:

- Faculty who collaborate with scholars in other countries, work on international standards projects, teach in summer programs abroad, or attend international conferences may be able to suggest contact persons or organizations interested in sponsoring a communication internship. U.S. communication faculty who have advised students interning or traveling abroad include Gerald Alred, University of Wisconsin/Milwaukee; TyAnna Herrington, Georgia Institute of Technology; Cindy Nahrwold, University of Arkansas at Little Rock; Barry Thatcher, New Mexico State University; Thomas Warren, Oklahoma State University; and Carolyn Washburne, University of Wisconsin/Milwaukee. Other campus departments with international internship or guest faculty programs may provide useful contacts. While international employers in marketing, agriculture, finance, or the sciences may have limited exposure to technical communication as an academic discipline, they may be able to provide contact names in industry or other useful information if communication students explain clearly what they can contribute to an organization.

- Technical communication organizations with international memberships may offer leads. For example, INTECOM, organized in 1970, offers international networking and resources for technical communicators, with members in Europe, New Zealand, and North America. TCeurope (www.tceurope.org) is an umbrella group for technical communication organizations in 10 European countries. Within the Society for Technical Communication (www.stc.org)—the largest professional organization for technical communicators—the Web site for the Special Interest Group on International Technical Communication (www.stcsig.org/itc) lists links to organizations with employment opportunities outside the United States.
- More generally, the Web sites for Intern Abroad (www.internabroad.com) and TransitionsAbroad.com (www.transitionsabroad.com) are online clearinghouses for educational and student employment opportunities in a wide variety of countries and academic disciplines, including communications.
- Conference proceedings and professional literature suggest numerous subject matter experts in international organizations who could be good resources. For example, the Web site for the Association of Teachers of Technical Writing (www.attw.org) archives abstracts and lists of speakers at past ATTW conferences on international communication and teaching topics.
- Listservs and online special interest groups include many contributors who work in technical communication outside the United States. The STC Special Interest Group on International Technical Communication (www.stcsig.org/itc) maintains an active listserv. TECHWR-L (www.techwr-l.com) is another technical communication listserv with high traffic, members in numerous specialties and locations, and archived posts.

An academic program increases the likelihood of successful internships by defining the content areas in which potential interns are best prepared for projects and by using career testing, counseling, and behavioral interviewing to become well-acquainted with its students individually (Rosenquist & Staples, 1994). Once students and advisors have identified potential employers, interns research the organizations and host countries so that the résumés and proposals students submit will be appropriate in form and content. International studies offices on campus will have helpful information about prospective host countries and regulations governing short-term employment, visas, and entry for business purposes, which vary widely from country to country. The U.S. State Department Web site (www.state.gov) contains links to consular information sheets for individual countries.

In this stage, students also determine what their budgets will allow realistically for travel and living expenses. The cost of living varies with location and fluctuates with the exchange rate, but an internship abroad is likely to be more expensive than local employment.

In addition, students clarify the lines of communication, required assignments, and other paperwork within the home program or campus. They also research the support systems available while they are abroad (Mendenhall & Oddou, 1988). For example, I was able to look forward to on-site visits from my American advisor and family members, regular e-mails from home, and participation in a local religious group while I was working in Europe. Students need to know they will have consistent contact from the home culture, including regular e-mails with program faculty.

Communicating with Employers

In the expectation stage, the interns' most focused opportunities for contextual reading occur during formal and informal interactions with employers, as they assess each other and plan for the work ahead. The interns begin to read the letters, calls, and e-mails from abroad not only for their literal content but also for what they communicate about organizations and ways of doing things in different cultures. Through observation and reflection, interns may be able to answer the following questions about employers:

- Does the organization have a Web site? If so, what groups appear to be the company's target audiences?
- What products and services are emphasized?
- Does the Web site include a list of clients?
- What problems has the company addressed for those clients?
- How has the company positioned itself internationally?
- Does the company's public face match what the student has learned about the organization through other channels?
- What are the relative roles of English and other languages in the organization?
- How responsive is the organization's representative to the student's questions?

By reading contextually, students evaluate the fit between their skills and the organizations' needs, including tasks to be accomplished, language preferences, the length of each internship, and employment and travel laws in the host country. Initial communications are likely to extend over several months. In one empirical study, Mumford (2000) points out the importance of appropriate work placement for students in short-term arrangements and also suggests that assistance from an on-site cultural interpreter at intervals after arrival may be more helpful than orientation at the outset. The study, which focused on British youth (ages 17 to 19) doing volunteer work abroad, found that "problems at work are the major risk factor for culture shock" (p. 84), after issues of cultural distance such as differences in food, dress, climate, language, leisure, and the like. Students and advisors take the time necessary to ensure a good match between intern and employer.

If early interactions go smoothly, a student may ask whether the employer will permit an advance look at company materials. Several months before I actually went to work, the host company sent me intranet links to several of its manuals. I was able to begin thinking about the company's software products well ahead of time, which helped me to be productive more quickly when the internship began. I was also able to evaluate the levels of edit needed so that the English-language documentation would be acceptable to American readers, one of the tasks the company was proposing for the internship. This preliminary look at materials provided me with reading practice in two ways: a literal reading of text to be revised, and an early attempt at naming and framing the issues the company intended to address with its external audiences.

As soon as programs and companies have agreed on the basics of the internships, students should consider submitting formal proposals to the employers, whether the students' academic programs require this step or not. A proposal underscores a student's professionalism. It also summarizes important details of the internship for all parties and helps to prevent later misunderstandings. For example, although some employers may cover lodging or travel, the internships may otherwise be unpaid. Students and employers should communicate clearly on the issues of compensation, living expenses, and work permits in the host countries, if required. Proposals should also note whether employers must submit paperwork to the academic programs, such as signed contracts or post-internship evaluations, and how employers can obtain university forms and additional information.

Table 2 shows how topics in an internship proposal might incorporate points of view of employer and student.

I found that writing a proposal clarified the objectives of the internship for me, particularly concerning the information audit of the manuals I was to complete. The proposal made me aware of some important gaps in my knowledge of audits and information models while I still had time to do additional background reading on those topics. Through writing the proposal, I began the

Table 2. Sample Topics for an Internship Proposal

Proposal topics	Employer's perspective	Intern's perspective
Objective	Expectations for the internship	Deliverables to be completed
Schedule	Dates of the internship, total hours to be worked	Dates of the internship, total hours to be worked; activities excluded from on-the-job hours, such as writing assignments for faculty advisor
Methodology	Equipment to be provided	Procedures for completing tasks
Costs	Compensation or reimbursement	Compensation or reimbursement; request for help in arranging lodging, if applicable
University requirements	Documents to be returned to the university	Documents to be forwarded from the university to the employer; instructions and other information, as needed
Legal issues	Permitting process for temporary employment initiated, if necessary	Permitting process for temporary employment completed, if necessary
Intern's credentials	Relevant coursework or employment experience	Relevant coursework or employment experience

internship with greater confidence in my ability to produce deliverables that would be useful to the company.

In the expectation stage, students are likely to regard the internship as a vehicle for knowledge transfer and honing of skills, with some exciting foreign travel thrown in. However, reading up on the host country will not suffice as preparation. The critical task for students in this stage is to practice their overall reading of social environments so they can learn to balance cultural generalizations brought from home with the specifics they will observe firsthand.

STAGE TWO: DISORIENTATION

After months of anticipation, students reach the second stage of the internship: *disorientation* or *breaking-in,* the period in which they arrive in the host country and begin to sort out the new environment. Companies may offer formal introductions, and the students' supervisors may provide advice and rhetorical models to follow. In my case, on day three, my supervisor invited me to sit in on a software demonstration for a prospective client in another city. This session gave me a good overview of my company's product and allowed me to hear client questions relevant to the information audit I was to complete.

However, whether an organization is explicit about its expectations or not, interns begin reading contextually in earnest now. In contrast to neat textbook cases, the real work is likely to seem messy, ambiguous, fluid, and full of extraneous noise. For example, in the weeks since I had read the online documentation from the comfort of my home computer, the company had subdivided and added to the manuals. My earlier notions of content were now completely outdated. Further, I learned quickly that the company's project managers disagreed sharply on how to document the product and were somewhat skeptical of my ability to contribute much. Paper sizes, computer keyboards, photocopying, and telephone practices were all different from my U.S. experiences. Resources I had always counted on in American offices were no longer available, and I felt not only ill-prepared and unsure of my behavior, but less able to pick up the social cues around me (Pires, Stanton, & Ostenfeld, 2006). For new interns abroad, reading workplace situations can suddenly become quite complicated.

Away from the office as well, new interns must read the culture for most of their waking hours. At first, basic tasks such as buying food and traveling about in a new city without getting lost consume considerable time and energy. More than once, I took the wrong bus, missed my tram stop, or traveled many blocks out of my way while mapping out routes. Daily life sometimes resembles a laboratory experiment: If this, then . . . what? Sullivan and Porter's (1993) comment on research generally—the difficulty in knowing whether a particular observation is typical—is apt both during and after work, for interns may feel overwhelmed by unfamiliar details, yet unsure how to abstract from them any guiding principles. Praxis is not yet possible.

Consequently, major tasks for students in the disorientation stage include examining the assumptions they have brought to the new environment, listening actively at all times, and trying to understand communication goals from their colleagues' different perspectives.

Examining Assumptions

By definition, the disorientation stage challenges many of the students' professional and cultural givens. Interns realize that many North American communication and business practices are culture-specific, not universal. In the northern

Europe culture where I was working, for example, smokers and nonsmokers shared offices, which were generally smaller and less private than American professionals are accustomed to. The president of the company routinely made room in his own office for employee meetings. My co-workers tended to assume that clients and vendors would be multilingual (and were quite surprised that I was not). Employees seldom worked past stated business hours.

When new interns encounter cultural differences such as these, they may react defensively, protecting their worldviews and cultural preferences (J. Bennett, 1998). Primed with theories from coursework and American workplace experience, they may become overeager to demonstrate their value to the hiring organization. For example, they may be a bit too quick to play the part of English language expert by explaining native-speaker spellings and pronunciations. Conversely, they may be anxious to demonstrate their newly acquired cultural information and coping skills by offering unsolicited opinions on local customs, politics, and foods.

Yet reading situations for useful links and relationships is now critical, for "the greatest mistake a well-trained, well-educated, well-prepared, enthusiastic intern can make is to enter another country with the idea that as an American, the intern/guest automatically knows the 'right' way to accomplish whatever task is the goal" (T. Harrington, personal communication, December 19, 2004). A naïve, culturally inappropriate application of classroom theory to workplace practice will serve both intern and employer poorly.

Therefore, to balance the theoretical knowledge they have brought to the internship, students now begin to develop a practical understanding of the host organizations' standards and ways of doing things (Lutz, 1989). They start to learn the organizations' values, which can serve as "a heuristic for the writer's work" and help the interns understand "where, when, and how to argue for change" (p. 129). For example, the information model my European supervisor asked me to use from day one was a paradigm the program faculty in my home university had advised against. Classroom theory had prepared me for a different approach. However, by adopting the employer's model, I avoided an unnecessary challenge to my supervisor and built some credibility for changes I suggested at the end of the internship. Like researchers, interns are trying "to make sense of informants' accounts and explanations of everyday activities in the organization" (Jorgenson, 1999, p. 294). Students thus begin to read workplace issues realistically, not as they may have imagined the issues previously. Instead of assuming that they know what is happening, they observe and reflect, paying attention to their own emerging knowledge and discarding misconceptions in order to name and frame problems appropriately. I realized with some relief that I had read the information model issue correctly when a client visited the office several weeks later and noted approvingly that I was following the same model used in her own organization.

One of the best ways to build contextual reading skills is through interpersonal exchanges (Dias, Freedman, Medway, & Paré, 1999), though such socialization may be unfamiliar to students and therefore sometimes uncomfortable or unsatisfying to them (Lutz, 1989). Students may experience several types of social learning during their internships. At a basic level, the interns learn through "individual cognition" mediated by social factors (Dias et al., 1999, p. 185). This type of contextual knowledge is rudimentary, such as learning the roles in the writing process, the function of document cycling in the organization, and job-specific professional knowledge (MacKinnon, 1993). In my internship, the product demonstration on day three exemplified this basic type of learning. Up to that time, concepts in the host company's manuals had been abstract ideas for me, because I had no firsthand knowledge of the product. But after the demonstration, I could link concepts in the manuals with actual clients who sat across the table from me, describe their writing and translation tasks, and ask my company's presenters some of the same questions about the product that had occurred to me earlier. In addition, from the interaction between presenters and clients, I could see the importance of framing technical issues in ways meaningful to particular audiences. System administrators and end users asked different types of questions and needed different types of answers. During the rest of the internship, this demonstration served as a mental reference point for my writing, for I thought back to the persons around that conference table whenever I worked on the manuals. I was able to measure my classroom assumptions against my experience with actual audiences. Learning was individual, yet influenced by social elements.

As students build contextual reading skills, they learn collaboratively as well as individually. From working with others to reach shared understandings, students learn to read the place of texts within the company's workflow and other processes, and the company's rhetorical conventions (MacKinnon, 1993). They learn that feedback from superiors can signal important issues or distinctions within the company culture because, unlike school writing, workplace writing is epistemic and instrumental (Dias et al., 1999). I experienced this type of learning each time the company's project managers and I discussed (and sometimes debated) the best ways to address audience documentation needs—how readers might search for topics within the manuals, for example, or how best to categorize topics by users. I brought to the internship certain academic concepts of audience analysis, which the managers did consider; however, after I heard the managers describe the company's programming practices, I learned to balance documentation theory against the constraints and pressures of product development cycles. Interns learn about workplace context through interactions with others.

In the most advanced type of social learning, interns work directly with a skilled practitioner or participate in authentic tasks in a community of practitioners (Dias et al., 1999), building contextual knowledge of the organization as a discourse community that serves specific functions (MacKinnon, 1993). As interns become more adept at reading the workplace context, they realize that different types of writing from different sources are integrated to advance the organization's purposes. Before my internship, I had not fully considered the iterative aspects of a company's documentation, thinking of it rather as a one-way channel for information. I thought the company's programmers would simply create software and develop documentation, and the documentation would tell clients what they needed to know about using the software. In time, I realized that although the managers did frequently explain to clients how the product worked, they also listened to clients describing how *people* worked, which, in turn, affected new product features. For example, the programmers might customize certain software modules to provide a better fit with a client's internal approval chains. Thus the company's documentation was continually evolving to allow for a range of changing perspectives. The opportunity to work within an authentic community helped me to examine my erroneous assumptions.

Listening Actively to Audiences

In many organizations, interns occupy a unique position with respect to different audiences. Because the students may have been hired to support a particular service or product, they must respond to the needs of external customers. At the same time, because they are learners and subordinates, the students must respond to internal audiences as well. By their very presence in the organization, students affect the company dynamic, changing the context for the permanent employees (Anson & Forsberg, 1990) and compounding the fluidity the others already experience each day (Harrison, 1987). As a result, like ethnographic researchers, interns must gain entry and acceptance to be productive. A willingness to listen actively to colleagues and clients, consider alternatives, and change course, if necessary, is therefore essential for interns.

Active listening may mean continuing to question assumptions. For example, one widely accepted theory of cultural differences (Hall, 1989) holds that in low-context cultures such as the United States, technical documentation spells out processes explicitly for the audience, because readers expect to follow writers through a procedure step by step. By contrast, in high-context cultures, readers expect to figure out more of a given process for themselves, and writers are therefore explicit only about the situation for the procedure. Interns familiar with this theory may try to categorize their international workplaces accordingly. However, students may discover that similar divisions exist within disciplinary or professional subcultures, regardless of country. In my host organization, project managers from the same social, geographic, and professional culture held decidedly different views on the type, amount, and arrangement of information appropriate for external readers. Some of the project managers believed that documentation should provide only the basic information clients needed to carry out a procedure; others argued that clients needed product development material, programming theory, and marketing information in order to make sense of the documentation. High-context and low-context perspectives were thus represented within a single management group. Because the managers comprised my internal audience, I had to develop "a perspective-taking skill" (Jorgenson, 1999, p. 303) in order to produce an information model that was neither too general nor too narrowly situated for external readers. Active listening is essential for interns if they are to avoid being sidetracked by inappropriate paradigms or preconceptions.

Active listening is especially critical in the disorientation stage, when interns have lost so many of their familiar frames of reference. Interns who can listen with an open mind will experience fewer effects of culture shock because they are approaching the new environment in a positive way.

Accepting Difference Gracefully

Though international internships can be life-changing events for the students themselves, they should be realistic about the scope of their influence within host organizations. In the breaking-in stage, permanent employees likely have different expectations for the internship, which supervisors may not have outlined in advance. Degrees of acceptance vary among employees as well. Interns therefore acknowledge their limitations and focus on what is feasible for short-term employment. In my organization, some of the documentation disagreements had more to do with workflow and competing organizational priorities than with writing *per se.* Such issues were clearly outside the scope of the internship. Though interns may ultimately read the workplace context with some accuracy, they realize that longstanding conflicts within organizations will not be resolved solely as a result of their work.

Interns ease their acceptance among new colleagues by handling professional and cultural differences tactfully. They avoid extreme reactions to the frustrations of culture shock, neither rejecting the new environment nor idealizing the home culture (Oberg, 1960). According to one qualitative study of seasoned expatriates (Suutari & Brewster, 1998), cultural awareness should take precedence initially over technical skills. In this study, experienced employees advised newcomers first to integrate into the community, be receptive to the new culture, and take the adjustment process in stride, rather than emphasize their professional expertise.

Although they will never be cultural insiders, students strive to respect differences and participate appropriately in the second culture. For example, because the permanent employees may conduct routine business in languages the interns do not speak, interns observe office practices closely and ask others to clarify procedures they do not understand. Students also interact professionally with all employees, without jumping to conclusions about the office hierarchy.

They follow the conventions of daily office life, such as break times and smoking policies, use of telephones and photocopiers, dress standards, conversational topics, arrival and departure times, and attitudes toward personal space, though all of these may differ from common North American practices. They also show genuine interest in their hosts' culture and language.

Away from the office, students may notice additional cultural practices that differ from those of North Americans, such as driving habits, conventions for tipping, public transportation, use of air conditioning, and retail hours. In the disorientation stage, students' "analytic processes are often in high gear, drawing on an unlimited wealth of diversity for comparison and contrast" (J. Bennett, 1998, p. 222). Because guidebooks usually target tourists, not residents (even temporary ones), interns must be resourceful when it comes to figuring out shopping, household management, and other chores in a new culture. However, honest questions can spark interesting conversations, such as the time the receptionist in my office taught me about "bread in the wall," the coin-operated, streetside display cases of fresh bakery items. Cultural contrasts present challenges during an international internship. But as interns learn to read contexts appropriately, differences can also become great pleasures.

STAGE THREE: RESOLUTION

Eventually, in the *resolution* or *acceptance* stage, the interns are established in their roles. Though their positions are temporary, they achieve a measure of insider status, with some success in naming different workplace realities, framing them in context, and using those readings to respond appropriately to internal and external audience needs. Depending on the size of the organizations, the interns may have little direct contact with executive-level decision-making, but as their reading ability improves, they gain a greater sense of what MacKinnon (1993) calls "the less immediate, though rhetorically important, high-level context" (p. 54). In this stage, a client's visit to the office or the prospect of a new account for the company becomes an important event for interns and permanent employees alike. The interns feel a sense of ownership in the company's future. They want client interactions and product demonstrations to succeed. With greater understanding of the social environment come greater confidence and composing skill.

How are the interns learning these things? MacKinnon suggests the following strategies:

- Reading texts of others and writing their own
- Receiving feedback, including document cycling
- Developing subject-matter expertise
- Accepting supervision as employees
- Developing good listening and interpersonal skills

The interns' contextual reading becomes the means by which they achieve praxis, the creative tension between the classroom theory they brought to their internships and the workplace applications they have learned. From coursework, for example, I had learned (in the abstract) the value of analyzing carefully an audience's information needs. However, in my host company, I learned quickly that although the product managers were very attuned to client needs during product development, they seldom asked clients what they needed or wanted in documentation. Their approach was reactive rather than proactive, and I floated the idea of conducting documentation needs assessments a few times without success. Then one afternoon I heard one manager say to another (concerning a thorny documentation problem), "Let's just ask the customer." A few days later, a client visiting the office surprised us all by actually volunteering her team for usability testing of the documentation. Consequently, I saw firsthand how theory and application could be balanced in a professional setting when people have sufficient time and latitude to consider a new idea. To watch the emergence of praxis in the workplace is one of the most valuable experiences an intern can take home.

Outside the office, the students also experience growing personal satisfaction from reading accurately the contexts of daily life. They resolve some of the contradictions they puzzled over earlier, feel less defensive, and have a more flexible perspective on the host culture (J. Bennett, 1998). Although they may not quite copy the behavior they see around them, they have a better sense of what constitutes appropriate behavior in the host culture and how to repeat it for good results (Pires et al., 2006). Successes in wayfinding, shopping, community interactions, and weekend excursions reinforce a sense of modest cultural competence. At a higher level, students' improved reading skills help them to distinguish between what theory labels significant and generalizable (Sullivan & Porter, 1993) about the host culture with what practice reveals to be situated, contingent upon the behavior of individuals or specific custom. As the interns learn to think empathetically, they are able to temper the guidebook's opinion that locals are standoffish with the real-life, daily experience of the friendly neighborhood clerk. On some days, the interns actually feel more like residents than visitors.

HOW TECHNICAL COMMUNICATION FACULTY CAN HELP

The ability to read unfamiliar audiences and situations is partly a life skill that develops with maturity and experience. However, by teaching reflective thinking, cross-cultural case studies, organizational analysis, and authentic tasks, technical communication programs can strengthen students' contextual reading skills before internships begin, thereby helping to reduce culture shock and improve interns' ability to develop professional and personal praxis when they are abroad.

(1) Teach Reflection-in-Action

The single most important way in which students can prepare for an international work experience is to train themselves to reflect in action; that is, to think critically about what they are doing while they can still adapt to a situation (Schon, 1987). Depending on need, professionals may sometimes apply rules directly, even automatically, and may also reason efficiently from rules in new situations. However, they also understand how to think beyond stated rules in gray areas without correct answers, a distinction Schon likens to the difference between technical mastery and true artistry. Lutz (1989) observes a parallel distinction between generic and situated writing. Consequently, to help students move from the level of "knowing" to that of "reflecting," instructors can provide prospective interns with opportunities to analyze the contextual differences in various types of materials. For example, instructors might consider a compositional approach described by Graff (2003) in which students use their own well-developed, nonacademic interests as the basis for learning to write "an academically rigorous treatment of those interests" (p. 226). Faculty might ask students to create two versions of instructions for an activity or hobby the students know well—one set for a beginner, and a second set for an intermediate user—and then analyze the rhetorical and contextual differences between the versions in an academic paper. By thoughtfully examining "street" knowledge in which they are already experts, prospective interns may gain insight into ways they might similarly gather, contextualize, and reflect on new information they encounter in other settings.

(2) Teach Authentic Cross-Cultural Case Studies

An effective case study teaches not only by illustration but also by involvement, drawing the student into the problem narrative as one who participates in balancing theory and practice (Stewart, 1991). When a case study's narrative is truncated, with the solution and conclusion left unstated, the student-reader begins to experience problem-solving behavior. The case study thereby "conveys the theoretical paradigm to new members of the theoretical community by telling a story that shows the paradigm in action. . . . The neophyte sees not only what problems look like, but also what problem-solvers look like" (p. 122), and the student has a way to rehearse naming and framing new issues. In a low-risk way, case studies help to socialize students into the community of skillful contextual readers (Lutz, 1989). For example, Bosley (2001) offers a variety of fictional and nonfictional narratives of cross-cultural communication problems for discussion, ranging from the radical differences in correspondence styles of American and Korean executives to the role of communication in the 1994 sinking of a European passenger ferry. In addition, *Intercom* magazine regularly features practical articles on global teams and projects in technical communication that can supplement case studies. While no case study can truly replicate an organization or a second culture, a well-written case can incorporate some of the same ambiguity, contradiction, and uncertainty an intern is likely to encounter in an international setting.

(3) Teach Organizational Analysis

The process of formally analyzing organizational culture clearly requires contextual reading. This exercise therefore helps prospective interns to develop such skills. How are priorities determined in Organization A? What are its stated and unstated goals? How does information flow, both formally and informally? How are conflicts resolved? As students learn to identify the written and unwritten codes, assumptions, and values that operate in organizational settings generally, they develop a framework for reading the specific contexts that will inform their writing during an internship. For example, in addition to rhetoric and design issues, much of the literature on the Three Mile Island, Challenger, and Columbia accidents traces the flow of information (or the lack thereof) through organizations. Studies of these situations can therefore help students to become more aware of competing priorities and organizational hierarchies within a group. Such analyses create "some capacity for examining social life within an organization as if it were a text . . . . [They also help students] both to participate within and to stand outside the prevailing systems and structures for assigning meaning within a given context" (Harrison, 1987, pp. 18–19), just as they will in their international workplaces. Adler (1997) expands the discussion of organizational behavior to include international issues such as ethical assumptions and acceptable work behavior in different cultures. Because students may be unsure of the issues to analyze, instruction should include specific strategies for gathering the information needed (Harrison, 1987; Lutz, 1989). As an introduction to organizational analysis, for example, students might examine the information flow or the ways decisions are made in a small campus organization or in their local workplaces.

(4) Teach Authentic Tasks

In contrast to textbook exercises, the work students will perform for employers is purposeful writing (English & Jereb, 1995). Consequently, when academic programs give prospective interns opportunities to complete authentic projects for real clients, they help students "experience the practitioner's rhetorical life and, in so doing, learn something about workplace writing" (Dias et al., 1999, p. 220). In a technical communication program, authentic tasks could include feasibility reports, editing projects, document redesigns, policy manuals, or other writing tasks in response to real-world situations. These purposeful projects provide a form of "scaffolding" (Dias et al., 1999, p. 221) that supports students as they attempt unfamiliar tasks, particularly when students can receive feedback and advice from experienced practitioners. Ideally, purposeful writing also exposes students to realistic, systematic audience analysis. With faculty guidance, for example, upper-division undergraduates can interview real-world clients concerning communication issues or problems. They can also set up and conduct small-scale usability testing of target readers for documents developed for assignments. Such tests give students (perhaps for the first time) some focused feedback from readers other than the instructor. Prospective interns need to see firsthand how and why real audiences respond to documents in particular ways, beyond superficial demographics or reader profiles. In addition, certain audience factors may have cross-cultural implications, and faculty can help students identify and analyze these elements. In working directly with real-world audiences on even a limited basis, students gain authentic points of comparison that will help them to discern contextual issues in the international workplace later on.

CONCLUSION

Some students make the transition from school to workplace writing easily. Whether it arises from inherent perceptiveness or well-developed social skills, the "rhetorical acumen" of these students enables them "to relearn writing when they switch from the epistemic social motives of academia to the instrumental social motives of the workplace" (Dias et al., 1999, p. 183). Regardless of location, these students begin internships with some contextual reading skills already in place. Many other interns, however, are preoccupied by the perceived need to demonstrate technical mastery right away. For these students, the idea of stepping back to reflect critically on assumptions and strategies produces discomfort. Further, they are unsure about the best ways to develop information resources within organizations, which has repercussions for the written texts they produce. When the internships are international, issues of culture shock compound students' uncertainties.

Academic programs that help their students to name unfamiliar professional and cultural phenomena and frame them in context thus provide those students with critical support, as well as an interpretive strategy essential in the international workplace. In learning to discern with care, students develop a foundation for differentiating between local, particularized phenomena with narrow application and those of general significance. They begin to balance *episteme* and *techne* more effectively, thereby achieving the praxis that produces creative solutions. More broadly, because international interns are immersed in a second culture, they become more comfortable with contextual reading and problem-solving skills they will use around the clock. At work, they learn to move more confidently in the sometimes complementary, sometimes conflicting spaces of communication theory and practice. Everywhere else, they learn to appreciate reading directly from a fascinating and unique life text.

Portions of this article were published in the May 2005 issue of *Intercom*.

REFERENCES

Adler, N. J. (1997). *International dimensions of organizational behavior* (3rd ed.). Cincinnati, OH: South-Western College Publishing.

Anson, C. M., & Forsberg, L. L. (1990). Moving beyond the academic community: Transitional stages in professional writing. *Written Communication, 7,* 200–231.

Barna, L. M. (1998). Stumbling blocks in intercultural communication. In M. J. Bennett (Ed.), *Basic concepts of intercultural communication* (pp. 173–189). Yarmouth, ME: Intercultural Press.

Bennett, J. M. (1998). Transition shock: Putting culture shock in perspective. In M. J. Bennett (Ed.), *Basic concepts of intercultural communication* (pp. 215–223). Yarmouth, ME: Intercultural Press.

Bennett, M. J. (1998). Overcoming the golden rule: Sympathy and empathy. In M. J. Bennett (Ed.), *Basic concepts of intercultural communication* (pp. 191–214). Yarmouth, ME: Intercultural Press.

Bosley, D. S. (2001). *Global contexts: Case studies in international technical communication.* Needham Heights, MA: Allyn & Bacon.

Campbell, C. P. (1998, February 20). *Beyond language: Cultural predispositions in business correspondence.* Presented at the Region 5 Society for Technical Communication, Conference, Ft. Worth, TX. Retrieved April 15, 2007, from http://infohost.nmt.edu/~cpc/internationalethos.html

Dias, P., Freedman, A., Medway, P., & Paré, A. (1999). *Worlds apart: Acting and writing in academic and workplace contexts.* Mahwah, NJ: Lawrence Erlbaum Associates.

English, K. A., & Jereb, B. (1995, April 23–26). *Communication and the internship: The roles of program directors, teachers, and corporate supervisors in facilitating change.* Proceedings of the Society for Technical Communication. Retrieved October 10, 2005, from http://www.stc.org/confproceed/ 1995/PDFs/PG6667.PDF

Eschbach, D. M., Parker, G. E., & Stoeberl, P. A. (2001). American repatriate employees' retrospective assessments of the effects of cross-cultural training on their adaptation to international assignments. *International Journal of Human Resource Management, 12,* 270–287. Retrieved April 5, 2007, from http://web.ebscohost.com.proxy.missouristate.edu/ehost/pdf?vid=8&hid=103&sid=e1d9109a-2653-43ce-ab30-0289e5879177%40sessionmgr102

Geertz, C. (1983). *Local knowledge.* New York: Basic Books.

Graff, G. (2003). *Clueless in academe.* New Haven, CT: Yale University Press.

Hall, E. T. (1989). *Beyond culture.* New York: Anchor Books.

Harrison, T. M. (1987). Frameworks for the study of writing in organizational contexts. *Written Communication, 4,* 3–23.

Hofstede, G. H. (1991). *Cultures and organizations: Software of the mind.* London: McGraw-Hill.

Hunsinger, R. P. (2006). Culture and cultural identity in intercultural technical communication. *Technical Communication Quarterly, 15,* 31–48.

Jorgenson, J. (1999). The student-as-researcher in international organizations. In C. R. Lovitt & D. Goswami (Eds.), *Exploring the rhetoric of international professional communication* (pp. 293–304). Amityville, NY: Baywood.

Lutz, J. A. (1989). Writers in organizations and how they learn the image: Theory, research, and implications. In C. B. Matalene (Ed.), *Worlds of writing: Teaching and learning in discourse communities of work* (pp. 113–135). New York: Random House.

MacKinnon, J. (1993). Becoming a rhetor: Developing writing ability in a mature, writing-intensive organization. In R. Spilka (Ed.), *Writing in the workplace: New research perspectives* (pp. 41–55). Carbondale, IL: Southern Illinois University Press.

Mendenhall, M. E., & Oddou, G. (1988). The overseas assignment: A practical look. *Business Horizons, 31,* 78–84. Retrieved April 5, 2007, from http://web.ebscohost.com.proxy.missouristate.edu/ehost/pdf?vid=4&hid=118&sid=52ab1452-1d70-4775-9d2e-02837427f1f0%40sessionmgr106

Mumford, D. B. (2000). Culture shock among young British volunteers working abroad: Predictors, risk factors and outcome. *Transcultural Psychiatry, 37,* 73–87. Retrieved April 5, 2007, from tps.sagepub.com/cgi/content/abstract/37/1/73

Oberg, K. (1960). Cultural shock: Adjustment to new cultural environments. *Practical Anthropology, 7–8,* 177–182.

Pires, G., Stanton, J., & Ostenfeld, S. (2006). Improving expatriate adjustment and effectiveness in ethnically diverse countries: Marketing insights. *Cross Cultural Management, 13,* 156–170. Retrieved April 5, 2007, from http://www.emeraldinsight.com.proxy.missouristate.edu/Insight/viewPDF.jsp?Filename=html/Output/Published/EmeraldFullTextArticle/Pdf/1360130205.pdf

Rosenquist, D. J., & Staples, K. E. (1994, May 15–18). *Planning successful internships: Matchmaking for organizational culture.* Proceedings of the Society for Technical Communication. Retrieved October 10, 2005, from http://www.stc.org/confproceed/1994/PDFs/PG135136.PDF

Schon, D. A. (1987). *Educating the reflective practitioner.* San Francisco, CA: Jossey-Bass.

Scollon, R., & Scollon, S. W. (2001). *Intercultural communication: A discourse approach* (2nd ed.). Malden, MA: Blackwell Publishing.

Stewart, A. H. (1991). The role of narrative structure in the transfer of ideas: The case study and management theory. In C. Bazerman & J. Paradis (Eds.), *Textual dynamics of the professions* (pp. 120–144). Madison,: WI University of Wisconsin Press.

Sullivan, P., & Porter, J. E. (1993). On theory, practice, and method. In R. Spilka (Ed.), *Writing in the workplace* (pp. 220–237). Carbondale, IL: Southern Illinois University Press.

Suutari, V., & Brewster, C. (1998). The adaptation of expatriates in Europe: Evidence from Finnish companies. *Personnel Review, 27,* 89–103. Retrieved April 5, 2007, from http://www.emeraldinsight.com.proxy.missouristate.edu/Insight/viewPDF.jsp?Filename=html/Output/Published/EmeraldFullTextArticle/Pdf/0140270201.pdf

Trompenaars, F. (1993). *Riding the waves of culture.* London: Nicholas Brealey Publishing.

Tufte, E. (1983). *The visual display of quantitative information.* Cheshire, CT: Graphics Press.

CHAPTER 11

ISO Standards and Cross-Cultural Communication: Materials for Teachers[1]

Thomas L. Warren

Standards are everywhere, from the size of your credit cards to the way computers systems exchange data to the size of the faucet opening (and threading) to how you find a book in the library. They allow commerce to proceed with minimum confusion and conflict. In addition, standards also represent examples of how cross-cultural communication is supposed to work, because standards are documents that are to be culturally neutral so as to be accessible and acceptable to users in any culture. Therefore, teachers can use standards as tools to teach students about cross-cultural communications in an industrial or corporate context.

[1] Throughout this chapter, I have focused on standards that students most likely will encounter in their professional employment (and perhaps become a part in producing). I will address both ISO standards related to technical communications and ISO standards related to other disciplines. I will draw on my experience working with ISO Technical Committee 184 (Industrial Automation Systems and Integration), Subcommittee 4 (Standardization of Information That is Shared or Exchanged in Industrial and Manufacturing; specifically, Product Data, Industrial Manufacturing Management Data, and Global Manufacturing Programming Languages). For those interested in ISO standards related to technical communication, the appropriate place to look is in the work of Technical Committee 10. See also the bibliography for links to TC10 standards. In addition, standards are a constantly changing entity, as are the supplemental documents associated with them. I have tried to provide the latest versions of all materials, but with the time between completed manuscript and the book in your hand being many months, I cannot always be sure that I have provided references to the latest versions. However, the most current versions of all documents I mention are not hard to find.

Many are mistaken about standards and the roles they play. When people hear that a certain item is designed according to a specific standard, they assume that the manufacturer had to follow the standard. Only some governmental situations mandate following standards; for all others, they are optional and voluntary. Of course, if you want your debit card to work in an ATM in California and then in New York and then Berlin, it has to conform to the international standards used to design the ATM card and reader.

Because the committees that draft the standards and the organizations that approve the standards are multinational, the final products must conform to local laws and customs to receive approval. The problem is that the International Organization of Standardization (ISO) cannot localize standards to every culture the way companies do. When you read a standard—particularly an international standard from the ISO—you will not see words like *must*, as in, "You *must* do. . . ." The word does not appear in standards because if it did, then the ISO standard could possibly be in conflict with local law, and that conflict would prevent the standard from becoming international. Instead, *should* and other such words appear.

An interesting part of the standards community is that it is an amalgamated culture; that is, it has a culture that does not exist apart from the specific community itself and is made up of multiple cultures. Persons from different cultures come together to work on a standard and during the time they spend together, they form a temporary culture to prepare the international standard. When the members are not working on their standard, the new culture does not exist in either a subjective or objective form (Limaye & Victor, 1991). Developing communications—drafts of standards, reports on standards development, proposals for new standards, and such—offer significant cross-cultural communication challenges not normally found in other such contexts. While you can have multiple cultures represented at, for example, an international conference or meeting such as an INTECOM's FORUM Conference, it is rare that the cultural aspects extend beyond that particular conference or meeting. The standards environment, however, can extend for a long time.

As you will see, standards can take up to 48 months to complete and be accepted. There are also the reviews and revisions that the same people who produced the drafts must manage. These challenges in creating standards are especially interesting because the technology that forms the content of the standard is generally culturally neutral. Whereas a human language may have considerable cultural components, a computer language such as EXPRESS typically is culture-free. Discussions about cultural implications of a technology rarely happen as the group discusses and drafts the various versions; however, culture does play a part in discussing matters such as language (more about that later) and social issues.

Standards have been around for a long time. There is the story about how roads in Rome and elsewhere were constructed based on the width of a Roman horse's rear end. This measurement led wagon and chariot makers to "standardize" on the width of their wagons and chariots. And that led to the width of railroad tracks and on and on. Nowhere in the story do you find any group (governmental or industrial) forcing such measurements on people. Rather, the manufacturers voluntarily accept the measurements because they wanted their products to "fit" on the roads.

Roads are not the only areas where a standard evolved based on economic interests. Here are some other examples:

- The United States is the only major country in the world to use 8½ × 11 paper. The rest of the world uses 210 × 297mm (A4—ISO 216).
- The United States does not use metric measurements (SI Units—ISO 1000).
- Considerable problems ensued in the 19th century because railroad tracks were not the same width. Freight had to be unloaded and reloaded when the shipper changed from one country's train system to another.
- When the German rail system had a crash near Hannover in the late 1990s, they could not borrow cars from France because of differences in widths (the French cars could not fit through the German tunnels).
- English follows standardized spelling (18th century) and grammar (18th century).
- Screw threads (ISO 68), without which hardware stores would have to stock hundreds or thousands of different screws.
- Building safety features (ISO 12543)
- SGML (ISO 8889)
- XML (ISO 8825)

The need for standards arises when there are conflicts in product specifications. For example, in the 20th and 21st centuries, we find competing formats battling in the market place for survival: BetaMax vs. VHS; the 8-track tape vs. the cassette; and more recently, HD DVD vs. Blu-Ray. Eventually, the market determines which specifications become a standard. So one way standards appear is through the marketplace, and culture could play a major role in that process but rarely in how standards are produced. I want to focus then, on the way standards come about. Standards happen through a multinational consensus process that reflects governmental and industrial needs and involves considerable cross-cultural communication.

This chapter provides teachers with the background needed to understand how a standard evolves and how cross-cultural communication plays a part in developing the final approved version. I will first explain how an international standard is developed, moving from a Preliminary New Work Item (PNWI) to the completed and accepted International Standard (IS). While the multicultural makeup of the committees is an important part of this process, the dominant process that overrides even cultural considerations (and is actually a part of them) is consensus, because all standards activity comes through consensus. Second, I will discuss how teachers can use standards and standards development to teach their students about cross-cultural communication issues.

DEVELOPING STANDARDS

What is a standard? Standards are agreements that are documented by precise criteria among nations. As agreements, they are treated as suggestions, guidelines, or definitions that ensure that materials, products, or services are fit for their purposes. Industries, through their national standards bodies, identify their country's or industries' needs and provide (and pay for) volunteers to develop standards based on their interests. All aspects of government, industry, business, and academia (not to mention consumers) have vested interests in seeing standards accepted internationally so that better products and services motivate organizations to participate in standards work.

How are Standards Developed and Adopted?

There are over 15,000 standards currently. These standards are developed by groups of dedicated people, working together in a cross-cultural communication environment. Each country has at least one national standards body that is supposed to be nongovernmental. In the United States, that body is the American National Standards Institute (ANSI);[2] in Great Britain, British Standards Institute (BSI); in China, China Standardization and Information Institute; in Japan, Japanese International Standards (JIS); and so on. These national bodies collectively belong to an international body called the International Organization for Standards (ISO) through which international standards are developed. Membership in ISO is in three classes: Participant Members (P-Members who participate and vote), Observer Members (O-Members who participate but do not vote), and Liaison-Members (interested professional organizations and companies who usually do not participate and do not vote but do receive reports and communications).

Brief History of ISO

While standards have been around for a long time, an international standards organization established to coordinate the efforts is a 20th-century invention.

According to its Web site (http://www.iso.org/iso/about.htm), ISO evolved from another international organization, the International Electrotechnical

[2] ANSI, through NISO, has produced a national standard on technical and scientific reports (Z39-19—2006 [American National Standards Institute, 2005]).

Council (IEC), founded in 1906 and still functioning jointly on some standards with ISO. The first meeting of representatives that eventually formed ISO occurred in London in 1946. However, ISO's official birthday is February 23, 1947, and it approved its first standard in 1951: Standard Reference Temperature (International Organization for Standardization, 2002), revised several times. Since then, they have produced thousands of standards in many disciplines. Most are sold to support the central organization, but there are some that are freely distributed.[3]

The mission of ISO is to increase economic as well as scientific and technical exchange of products and services. To accomplish this mission, ISO members meet and develop the standards needed by the member nations and the industries within those nations.

The acronym, ISO, may seem a misreading, but it really is not an acronym, *per se*. The letters are derived from a Greek word *isos* meaning "equal"; for example, the isosceles triangle has equal sides, thus stressing the equality all interested parties have in the process. The international nature of ISO standards is shown in that all standards are published in three languages: English, French, and Russian. All committee work and meetings, however, are conducted in English. The principle language authority for drafting all standards is the *Concise Oxford English Dictionary*, 9th edition (Thompson, 1995), making the English used primarily British English.[4]

Organization

The participating national members (P-Members) nominate experts who will develop the standard. These P-Members, through the General Assembly and Council, maintain the administrative structure of the organization, headquartered in Geneva, Switzerland. ISO has a number of technical committees (TC) that are discipline-specific. Each technical committee has a number of subcommittees (SC) that work on different parts of the discipline. One such technical committee is TC 184, charged with developing standards in *Industrial automation systems and integration—Product data representation and exchange* with several subcommittees. TC 184/SC4, for example, develops standards relating to *The Standard for the Exchange of Product model data* (*STEP*—ISO 10303).

As an example, the P-Members for ISO TC 184/SC4 are listed in Appendix A. Notice the range of countries involved and the potential cross-cultural communication problems. It is also instructive to note that a large number of countries are not represented among the P-Members (see Appendix A), O-Members (see Appendix B), and Liaison Members (see Appendix C) for TC 184/SC4. Principally, most countries from the Middle East are absent. While it would be

[3] http://isotc.ios.org/livelink/livelink/fetch/2000/2489/Ittf_Home/PublicallyAvailableStandards.htm

[4] This requirement may have changed to a more recent edition of the *COED*.

interesting to speculate as to the reasons for such an absence and if their nonparticipation relates in any way to cultural issues, a more reasonable explanation is that they are not interested in the issues developed by TC 184/SC4 and so, do not participate in developing standards related to that aspect of industrialization.

Overall

ISO Geneva, as the central administrative unit is frequently called, develops policies and procedures that the committees follow. Chief among these policies are the *Directives*, which specify how the committees and subcommittees will develop the standards, thus functioning as the organizational style manual (International Organization for Standardization, 2008a).

Technical Committees

The technical committees (TC) are the heart of ISO. It is these 192 (currently) committees that do the work of the organization and develop the standards. The TCs are also where attention to cross-cultural communications is important. Each TC has one or more SC and each SC will have working groups (WG). Because many standards overlap, you will also find joint working groups (JWG) formed within SCs, across SCs, within TCs, and across TCs.

The technical committees as well as the subcommittees all have officers. At a minimum, the officers are the Chair and the Secretariat. They may have other officers should they choose.

P-Member organizations can appoint people to these TCs, SCs, JWGs, and JTCs. Because the work on standards is voluntary, those appointed to do the work must have funding from their employers because the ISO provides no funds for the expenses of those working on standards, and the national standards bodies frequently do not provide funds because of their limited budgets. A national standards body hosts standards meetings, where much of the work is done, three times a year (for TC 184/SC4) for one or more weeks, and they usually assess attendees a meeting fee to cover the cost of hosting the meetings. There are no registration fees, and all meetings are open to anyone wanting to attend and expressing a legitimate interest in standards development.[5] In addition, subgroups (typically WGs) will meet three or more times a year, usually before the SC meetings, to continue their work and develop the progress reports presented at the meetings.

[5] Those attending should understand that there are no sessions to "catch up" newcomers other than a general introductory session providing an overview of the TC or SC. Those attending are assumed to have expert technical-level knowledge in the subject of the WG's work.

Working Groups

The WG is where the work takes place. These groups consist of individuals representing at least five P-Member countries who must agree to work on the new standard, manage the voting and ballot comments made by the voters, and keep the standard current. Often, the SCs will have additional committees that do specialized work. ISO TC 184/SC4, for example, has

- A quality committee to ensure that the drafts conform to the ISO requirements;
- A change-management committee, which reviews preliminary new work items, new work items, and progress on all standards being developed by SC4; and
- The policy and procedures committee, which is responsible for oversight of the technical work.

As always, the membership of these committees, like that of the WGs, is made up of P-Members' appointees who are working on the SC, so the organizational structure of an SC can be complex. And cross-cultural communication issues are as evident in committees as in WGs.

Quality Committee

Because the TC 184/SC4 Quality Committee's role in developing standards for ISO TC 184/SC4 is so important, ensuring compliance with ISO's *Directives*, I want to explain how it works, especially with regard to cross-cultural communication issues. The quality committee (QC) ensures that TC 184/SC4 documents are of the appropriate quality, both in expression and meeting the ISO requirements. QC facilitates integrating quality in a holistic fashion throughout the life cycle of TC 184/SC4 drafts, using and tailoring existing ISO quality standards and definitions where applicable to define and disseminate repeatable, uniform processes that WGs can use consistently within TC 184/SC4 to produce high quality drafts.

To ensure that all drafts meet ISO requirements, QC developed two tools for the WGs to use:

1. A detailed explanation of how the standards are to be prepared. This document, the *Supplemental Directives* (TC 184/SC4 Quality Committee, 2001a), does not replace the *ISO Directives*, but, as its name suggests, supplements them.
2. A series of checklists for the WGs to follow to ensure compliance with ISO requirements (TC 184/SC4 Quality Committee, 2001b). The checklists are for

a. The WG leader (called a *WG Convenor*): The Convenor's Checklist (Quality Committee [document] Number QCN 195)
b. The WG Project leader: The Project Leader's Checklist (QCN194)
c. The WG members: The Internal Checklist (QCN 193).

The WG project team complete these checklists that then become part of the document package sent forward as a draft of the standard. In addition to these publications, QC maintains the *Quality Manual* (TC 184/SC4 Quality Committee, 2000) that details the quality processes each WG in SC4 should follow.

Developing a Standard

ISO procedures allow for a variety of standards documents:

- **An International Standard (IS—including amendments).** This document is normative; that is, it is a standard that industries should follow. It requires a two thirds vote of approval by the P-Members and less than one fourth negative vote. The standard is reviewed and revised every 5 years or when needed.
- **Technical Specification (TS).** Another normative document. It also requires two thirds vote of approval by P-Members. It is reviewed after 3 years with a maximum life of 6 years. After that, the WG can recommend that it become an IS, when it will follow the normal balloting procedures for an IS.
- **Publicly Accepted Specification (PAS).** This normative document requires a simple P-Member majority favoring it. It is reviewed also after 3 years and has a maximum life of 6 years. After 6 years, the WG can recommend that it become an IS, when it will follow normal balloting procedures.
- **Technical Report (TR).** This document is informative, meaning it is for information only and should not be considered normative. It requires a simple P-Member majority and has no review schedule.
- **International Workshop Agreement (IWA).** A normative document, this document requires approval only from the Technical Management Board—an advisory body to ISO-Geneva—and can become an IS later following normal balloting procedures.

Standards are approved following one of four paths. (As you read about these processes, remember that all groups involved in the procedures are multicultural):

1. **The normal path:** The work progresses through 6 stages:
   - A new work item (New Proposed Work Item) is proposed and becomes a Preliminary New Work Item (PNWI) if approved by the appropriate TC/SC. The WG then begins the technical work on the draft.

- The WG builds technical expert consensus through the draft document (Working Draft [WD]). If it meets all necessary criteria and receives the required number of "yes" votes within the WG, it moves to the next stage.
- The WG builds consensus within the SC and TC to produce the committee draft (CD). If it meets all necessary criteria and receives the required number of "yes" votes within the WG, TC, and SC, it moves to the next stage as a draft international standard (DIS).
- The SC conducts the first ballot of the DIS among the P-Members (note that previous votes were all within the TC, SC, and WG). If it meets all necessary criteria and receives the required number of "yes" votes, it moves to the next stage. Voters may elect to make comments regarding the DIS relating to any matters: style of writing, conforming to ISO *Directives*, content, and such. The WG then collects these comments and addresses each one. That document becomes a part of the standard's documentation and the DIS becomes a final draft (FDIS).
- The SC conducts a second ballot on the FDIS. If it meets all necessary criteria and receives the required number of "yes" votes, it becomes a standard.
- ISO Geneva then publishes it as an international standard (IS).

Note that the WG receives ballot comments regarding the draft and must respond to each comment. This process, called the Standard Enhancement and Discrepancy System process (SEDS), results in a document added to other documents (checklists, draft of the standard, etc.) in the group of draft documents (see Appendix E).

2. **Abbreviated paths**:
   a. An abbreviated path: The Preliminary New Work Item ballot moves the proposed standard to Committee Draft status or the draft is registered as a Draft International Standard (requires full agreement and completing a checkbox on a form).
   b. A second abbreviated path: The working draft moves to Draft International Standard with two thirds of the P-Members voting to approve.
   c. A third abbreviated path: The Draft International Standard bypasses Final Draft International Standard stage by receiving 100% approval at the Draft International Standard stage.
3. **An existing national standard**: P-Members and Liaison Members can propose an existing stable standard (such as a national standard) for consideration as an international standard. The fast track is to consider it as a Draft International Standard. It can also be combined with a New Work Item ballot.
4. **Revising an existing standard**: Revision can occur at any time and usually on a regular cycle of 3 to 5 years. The revision moves through the process following the normal path or any of the abbreviated paths.

The time required to create a standard can be

- Normal: 36 months usually, or
- Accelerated: 24 months (skips the Committee Draft stage), or
- Expanded: 48 months (expands the DIS and FDIS stages, usually because of technical concerns)

The WG Documents

New Work Item (NWI): Standards are developed by first having a new work item (NWI) proposed by a P-Member. In some TCs and SCs, there is a committee that will review the NWI and recommend its acceptance, rejection, or modification. The P-Member proposing the NWI must name at least five P-Member countries that agree to have experts work on the NWI. Documents that are generated for the New Work Item include:

- Preliminary Work Item (PWI): Outlines the need for a new standard.
- Proposal for a New Work Item (PWI): Proposes that ISO develop a new standard. The proposal is voted on by the P-Members and, if approved, signals that work can begin. Once approved by the P-Members, the working group forms and begins work on the new standard. The proposal is now known as a New Work Item (NWI).
- Working Drafts (WD): Working drafts of the proposed standard are prepared and reviewed by the internal organization of the SC of the TC.
- Committee Draft (CD): Once the draft has received consensus in the WG, it is ready for review by participating P-Members and interested industries. It is now called a CD. Following this draft, all technical work usually ends except for responses to the ballot comments. The focus shifts from the technical details to how the draft conforms to the ISO specifications including the communication issues. Because ISO has established the *Oxford Concise*, 9th edition (Thompson, 1995) as authoritative, few issues concerning vocabulary and spelling occur. Rather, issues concern format; consistency of expression; grammar, punctuation, and style; and clarity of expression. Here is where the influence of a culture could be seen but rarely is.

Proposal, Draft International Standard (DIS): When the CD achieves consensus, it is ready for review and balloting by ISO member countries. Voters may vote to approve, disapprove, abstain, or approve with comments. This ballot frequently generates a lot of voter comments that are returned to the SC and the WG for resolution using SEDS. The WG must address all comments with a formal response. A document containing the comments and the responses/resolution becomes part of the documentation attached to the DIS.

Ballot comments offer another occasion when cross-cultural communication issues could be important. Other than occasional errors in expression and punctuation, the comments are focused on the draft and the ISO *Directives* (in the case of TC 184/SC4 drafts, comments also relate to the *Supplemental Directives*). This focus, as with the WG's focus on technical matters, tends to bypass cultural issues in favor of format and wording issues.

Final Draft for Approval (FDIS): Following resolution of all ballot comments, the final draft is circulated to verify editing and responses to ballot comments. If approved, the FDIS is now ready for publication as an international standard.

International Standard (IS): Following the FDIS ballot and approval, the new international standard is then printed and distributed. It is also translated into French and Russian.

Cross-Cultural Communication Issues

Because the WG members come from different countries and the governing principle of standards work is consensus, potential cross-cultural communication issues are present at many levels. But because the work concerns technical matters, the WG rarely faces cross-cultural communication issues as it develops the technical drafts. That is not to say that there are no issues raised during the technical work and in ballot comments. What concerns me here are the cross-cultural issues that occur with communication. Yet, as I have shown (Warren, 2005), there can be many cross-cultural elements in composing and editing the drafts. Even the guidelines used by the working groups are not immune to language problems at both the technical levels and word levels. The *ISO Directives* and, in the case of TC 184/SC4, the Quality Committee's *Supplemental Directives*, focus on the form and format of the draft rather than on issues that draft editors should also be examining—style, rhetoric, cohesion, coherence, and so on—all subject to cultural influences. Even in ballot comments throughout the process, very few language comments come forward, and almost no cross-cultural problems appear.

To take one instance, analysts of cross-cultural communication can focus on the communities involved and the kind of language used within and between these communities. Few of the members of the working groups have a background in editing or writing because they are chosen for their technical expertise and not their facility with language. Yet the various guidelines assume that the WG members are skilled in language use. If we look at the kinds of language used in drafting standards, we find that it is heavily technical, as it should be. This technical language suggests Bernstein's Restricted Codes (Bernstein, 1971), because the standard is written for readers who are members of the technical community involved. Within the various guidelines themselves, it should be recognized that the editors are not members of the editing community (in this case, language specialists and editors) and should use Elaborated Codes. Yet the work of the standards community continues, and we get a steady flow of standards.

Using the definition adopted by the editors of this anthology (Hofstede, 1997), or almost any definition of *culture*, it is difficult to understand why there are not more problems with cross-cultural communication. Perhaps it is because the WG members put away cultural (and political) views when developing a standard. It is certainly true that consensus would be difficult to reach if members were focusing on cultural communication issues as well as the technical issues.

Or perhaps the answer lies in the view that technology is normally unaffected by culture. Cultures may encourage or deprecate certain technologies, but that does not change the neutrality of the technology. Likewise, judgments about technology being good or bad typically rest on how the technology is applied rather than anything to do with the technology itself. Therefore, WG members rarely have to confront cultural issues relating to the technology as they work on the various drafts.

Cross-Cultural Issues in TC and SC Work

Just about every aspect of an ISO gathering reflects concerns for the cultures of those attending, from the obvious point of scheduling to avoid important religious holidays to the meals that meet special dietary requirements. The spirit of cooperation and consensus fostered among the TCs, SCs, and WGs as well as the various joint committees precludes serious cultural conflicts on standards issues. But, as with any endeavor involving a mixture of cultures, there can be language issues [10]. TCs, SCs, and WGs normally have someone acting as an editor for the various drafts of the standards leading to the final version. These editors have requirements that they must ensure the drafts meet, starting with the *ISO Directives*. To help them edit their drafts to ensure conformity to ISO requirements, TC 184/SC4 QC developed checklists for the WG members to use. These checklists are as "technical" as are the specifications included in any draft. That is, there is little or no room for interpretation (see Appendix F). The draft either *meets* or *does not meet* the specification.

As an example, let us consider how the WG drafts definitions of key terms. *ISO Directives* states (International Organization for Standardization, 2008b; see Appendix C)

> **C.3.2 Layout**
> The preferred term (set in bold type in the printed publication) shall be placed on a new line, after its reference number, starting with a lower case letter except for any capital letters required by the normal written form in running text. The definition shall be placed on a new line, starting with a lower case letter, except for any capital letters required by the normal written form in running text, and shall not be followed by a full-stop.

> **2.4.1**
> **delamination**
> separation of two adjacent plies resulting from a lack of adhesion

Thus, we see that the definition has these parts:
Reference Number: *2.4.1*
Term being Defined: *delamination*
Definition: *separation of two adjacent plies resulting from a lack of adhesion*

The QC *Supplemental Directives* states the following about definitions:

4.2.5.1 General requirements for definitions
Each definition that is specified in a standard consists of the following:
– the term being defined;
– one or more synonyms for the term (optional);
– an abbreviation for the term (optional);
– the definition of the term.

If a term has an abbreviation, the term being defined shall be given in full and shall be followed by its abbreviation in parentheses, using the following format:

> *[ISO required]*
>
> **3.x.y**
> **<term> <abbreviation for term, in parentheses>**
> <definition>
>
> *[end required]*

NOTE: This is the only case in any ISO standard where the subclause number and the subclause title are placed on different lines.

The abbreviation shall also appear in the abbreviations subclause.

Definitions shall not end with a period.

The terms defined in this subclause shall not appear in the table of contents.

The terms defined in this subclause shall be listed in the index.

For definitions that have been extracted from another document that is not a standard, or have been modified from a definition that appears in another standard, a note shall follow the definition. This note shall have the following form: Adapted from <reference to the document> [n], where [n] is a reference to the bibliography (see 4.3.2).

For terms that have commonly used synonyms, the synonyms shall be listed after the term's definition. Each synonym shall be separated from its predecessor by a semicolon, using the following format:

[ISO required]

3.x.y

<term>; <synonym>
<definition>

[End required]

No variations are permitted in the format used for the definitions in the drafts and subsequent standards.

The way a culture or language group defines terms can vary considerably. For example, in Western culture, you may define a term in many ways: analogy, analysis and classification, cause and effect, comparison and contrast, description, etymology, exemplification, explication, figurative language, history of the subject, illustration, negation or elimination, operational definition, and stipulatory definition (Journet & Kling, 1984, pp. 9–10).

Yet in cultures where authorities prevail, authors usually appeal to those authorities for meanings of terms—even technical terms. Other cultures rely on the interaction of its members to determine the meaning of all terms. An issue might arise where members of these two cultures could be working on a draft and culturally disagree over how they would define terms. Both the *ISO Directives* and the QC *Supplemental Directives* for TC 184/SC4 solve this problem in three ways:

1. ISO has established a dictionary authority [9].
2. The *ISO Directives* permit use of definitions in previously published standards if they are appropriate to the content and context.
3. The different *Directives* have established a stipulative-type definition that is presented in a consistent format.

The WGs find similar directives to govern how they develop the various parts of the draft. The result is that potential cultural conflicts over how the drafting group defines a term are minimized through the stipulated format and method.

USING STANDARDS AS CROSS-CULTURAL COMMUNICATION

Teachers can use standards (and the process for creating a standard as well as the various documents that act as a style manual for developing standards) as starting points for discussing cross-cultural communication issues and how the standards communities have solved the problem. Because there are a lot of standards that are free and the directive documents are also free, teachers have a wide variety of documents to use. And they can use them in multiple ways because the entire process of developing a standard is documented and freely available. Teachers can develop exercises based on standards that help the students understand how a WG avoids cross-cultural communication problems as it works to develop international standards.

Furthermore, standards are documents that are not localized in the usual sense of making them culturally neutral for the communities for whom they are intended. Standards are not aimed at the same context as other technical documents, and because they are aimed at the international communities, there is no local community that requires localization. ISO *Directives* stipulate, "Fitness for implementation as a regional or national standard: The content of a document published by ISO and IEC shall be drawn up in such a way as to facilitate its direct application and its adoption without change as a regional or national standard." [5; section 4.6] As such, students aspiring to work for multinational companies can find examples of how technical documents can succeed in spite of no specific localizing efforts.

TEACHING STUDENTS ABOUT STANDARDS AND CROSS-CULTURAL COMMUNICATION

Here are some sample exercises teachers can use with ISO standards and the documents used to create them:

- Select a standard and analyze it based on a particular culture. What are the cultural considerations for any document that will cross cultures? There is language, but what else? Consider especially the role technology plays in a culture.
- Download a copy of the *ISO Directives, Parts 1 and 2* and analyze for cultural elements.
- Analyze an ISO standard for writing strategies used by the authors that clearly reflect a concern for cross-cultural communication issues. Again, what are the cultural aspects of any document?
- What writing strategies did the authors of the *ISO Directives, Parts 1 and 2* use? Also analyze the *TC 184/SC4 Quality Committee Supplemental Directives*.

- Select a particular culture and examine the role standards play in that culture. Is the culture part of a political entity (country) that participates in ISO work? If not, then why do you think it does not participate?
- Go to a national standards organization Web site and download any free standard. What cultural elements would you expect to find in it? What evidence do you find that the authors considered cross-cultural communication issues?
- Look at the list of TC 184/SC4 P-Members (see Appendix A). Why are some countries not represented? (You might have a look at the kinds of technical work TC 184/SC4 does and then determine if that country has an interest or need in that subject.)
- Students with an interest in technical communication standards can find, for example, standards related to technical communication developed by a number of TCs (see Appendix D). Prepare a report to the class in which you list and briefly discuss the standards technical communicators must know when doing their jobs.

CONCLUSION AND RECOMMENDATIONS

Conclusion

As the list of P-Members for ISO TC184/SC4 shows, many countries are actively interested in standards and their development. Even the O-Members and Liaison Members (see Appendixes B and C) represent a widely diverse cultural group that has interests in how the standards are developed. Consequently, many countries are involved through WG experts. Yet in spite of differences that might be related to cultures, standards evolve and appear. It is hard to imagine another area where students can study how work is done in a multicultural environment to produce documents that minimize cross-cultural communication problems. Certainly the United Nations represents an environment where cross-cultural communication is at work. But because of both the political processes involved and how students fail to understand how UN work directly affects their lives, examples from the UN rarely make good studies. Not only are standards apolitical and acultural, but they also have a direct impact on the students, regardless of their majors. That is why they can be effective learning aids for when the students need to understand how cross-cultural communication works.

What of the future? What are the areas in which we will see new standards emerge? Some of the areas where future standards work will occur include

1. The environment
2. Service sector (personal financial services, market opinion, social research, tourism)
3. Security (ports, transportation, etc.)
4. Good managerial and organizational practices

5. Energy
6. Water supplies
7. Transportation

Perhaps some of your students will work to develop standards in these areas as experts or editors. It is certainly an opportunity to practice the lessons learned when studying cross-cultural communications.

Recommendations

1. Show your students how standards are developed as a document process with all that it involves, especially the cross-cultural communication elements.
2. Stress that over 15,000 international standards exist incorporating every discipline and study (if your students are studying to become professional communicators, point them to the Web sites for the TC groups involved in developing technical communication standards [see Appendix D]).
3. Have them develop lists of issues that have potential cross-cultural communication significance and then ask them to analyze standards in their fields as well as the requirement documents—especially the ISO *Directives*.

APPENDIX A: PARTICIPATING MEMBERS (P-MEMBERS) OF TC 184/SC4

Australia
Canada
China
Czech Republic
France
Germany
Italy
Japan
Korea, Rep. of
Netherlands
Norway
Portugal
Russia
Spain
Sweden
Switzerland
United Kingdom
United States

APPENDIX B: OBSERVER MEMBERS (O-MEMBERS) OF TC 184/SC4

Belgium
Brazil
Bulgaria
Denmark
Finland
Hong Kong, China
Hungary
Romania
Singapore
Yugoslavia

APPENDIX C: LIAISON MEMBERS OF TC 184/SC4

Currently, there are 56 Liaison Members. For a complete listing and details, see http://www.tc184-sc4.org/SC4%5FOpen/SC4%5FMembers/liaisons%5Fmembers/

APPENDIX D: ISO TECHNICAL COMMITTEES OF IMMEDIATE INTEREST TO TECHNICAL COMMUNICATORS

TC 10: Technical product documentation
http://www.iso.org/iso/en/stdsdevelopment/tc/tclist/TechnicalCommitteeDetailPage.TechnicalCommitteeDetail?COMMID=443

TC 37: Terminology and other language and content resources
http://www.iso.org/iso/en/stdsdevelopment/tc/tclist/TechnicalCommitteeDetailPage.TechnicalCommitteeDetail?COMMID=1459

TC 46: Information and documentation
http://www.iso.org/iso/en/stdsdevelopment/tc/tclist/TechnicalCommitteeDetailPage.TechnicalCommitteeDetail?COMMID=1757

TC 130: Graphic technology
http://www.iso.org/iso/en/stdsdevelopment/tc/tclist/TechnicalCommitteeDetailPage.TechnicalCommitteeDetail?COMMID=3376

TC 171: Document management applications
http://www.iso.org/iso/en/stdsdevelopment/tc/tclist/TechnicalCommitteeDetailPage.TechnicalCommitteeDetail?COMMID=4053

APPENDIX E: Sample SEDS Report

SEDS Status Report

SEDS #	*Standard #*	*Clause #*	*WG Lead*	*Start Date*	*Resolved*	*Status*	*Incorporated*	*Author*
1	10303-0203		WG 03	1994-11-29	1996-02-15	C		Haenisch, Jochen
	Problem: Description:	There seem to be missing two WHERE-rules for the manifold_surface_shape_representation that restrict the use of entity						
2	10303-0041	Figure F.20 on Page 176	WG 12	1995-03-14	1995-05-22	C	1999-03-24	Haenisch, Jochen
	Problem: Description:	The figure for the measure_schema on page 176 is wrong. It is actually the product_property_definition_schema.						
3	10303-0203		WG 03	1995-03-14	1996-02-15	C		Dhar, Shantanu
	Problem: Description:	There appears to be an error in WR3 of the entity (GEOMETRICALLY_BOUNDED_WIREFRAME_SHAPE_REPRESENTATION:						

```
WR3: SIZEOF (QUERY (gcs <* QUERY (it <* SELF/representation.items |
   'CONFIG_CONTROL_DESIGN.GEOMETRIC_CURVE_SET' IN TYPEOF (it)) |
   NOT (SIZEOF (QUERY (elem <* gcs\geometeric_set.elements. |
   NOT (valid_geometrically_bounded_wf_curve
   (elem, "CONFIG_CONTROL_DESIGN')))) = 0))) = 0;
```

The type of elem needs to be checked before applying valid_geometrically_bounded_wf_curve on it.

APPENDIX F: SAMPLE ITEMS FROM CHECKLIST

FROM CHECKLISTS

ITEMS IN LISTS

Requirements on lists are described in 4.5 and 5.3.2 of SD and 5.2.5. of IDP2.

YES	NO	NA	
d	d	d	1. Items in bulleted lists are preceded by a long dash (em dash). (*See 5.3.2 of the SD.*) Comments:
d	d	d	2. Bulleted lists are limited to four levels or less. QC recommends no more than 2 levels are to be used. (*See 5.3.2 of the SD.*) Comments:
d	d	d	3. Items in the first level of an ordered list are preceded by lower case letters, followed by a right parenthesis and a blank space. (*See 5.3.2 of the SD.*) Comments:
d	d	d	4. Items in the second level of a two-level or deeper ordered list are preceded by Arabic numerals, followed by a right parenthesis and a blank space. (*See 5.3.2 of the SD.*) Comments:
d	d	d	5. Items in the third level of a three-level ordered list are preceded by lower case Roman letters, followed by a right parenthesis and a blank space. Items in the fourth level of a four-level ordered list are preceded by upper case Roman letters, followed by a right parenthesis and a blank space. (*See 5.3.2 of the SD.*) **QC strongly discourages using three-level and higher-level lists.** Comments:

ACKNOWLEDGMENTS

A number of experienced members of the standards community have reviewed this essay and offered suggestions, correcting factual errors, and pointing out holes. Factual errors that remain, however, are mine and mine alone. For his help, I gratefully acknowledge the work of John Nazemetz, Associate Professor, School of Industrial Engineering and Management, Oklahoma State University, and a colleague on TC 184/SC4.

I also gratefully acknowledge funding provided for several years for me to work on the Quality Committee of TC 184/SC4 by the Computer-Assisted Transfer of Technology (CATT) project at Oklahoma State University.

REFERENCES

American National Standards Institute. (2005). *Z39.18-2005. Scientific and technical reports: Preparation, presentation and preservation.* Retrieved May 29, 2006, from http://www.niso.org/standards/standard_detail.cfm?std_id=801

Bernstein, B. (1971). *Class, codes, and control; (Vol. 1: Theoretical studies toward a sociology of language).* London: Routledge and Kegan Paul.

Hofstede, G. (1997). *Culture and organizations: Software of the mind.* New York: McGraw-Hill.

International Organization for Standardization. (2008a). *ISO/IEC Directives, Part 1, Procedures for the technical work* (5th ed.). Retrieved May 28, 2008, from www.iso.org/sdis/directives

International Organization for Standardization. (2008b). *ISO Directives, Parts 1 and 2. ISO/IEC Directives, Part 2, Rules for the structure and drafting of International Standards.* Retrieved May 28, 2008, from www.iso.org/sdis/directives

International Organization for Standardization. (2002). *ISO 1-2002.* Retrieved May 29, 2005, from http://www.iso.org/

Journet, D., & Kling, J. L. (1984). *Readings for technical writers.* Dallas, TX: Scott Foresman.

Limaye, M. R., & Victor, D. A. (1991). Cross-cultural business communication research: State of the art and hypotheses for the 1990s. *The Journal of Business Communication, 28,* 277–299.

TC 184/SC4 Quality Committee. (2000). *SC4 Quality Manual (SC4 N1110).* Retrieved May 28, 2008, from http://www.tc184-sc4.org/SC4_Open/SC4_Standards_Developers_Info/documentation.cfm?CFID=61899&CFTOKEN=23621008&FID=1724#1724

TC 184/SC4 Quality Committee. (2001a). *SC4 Supplementary directives: Rules for the structure and drafting of SC4 standards for industrial data. QCN 200.* Retrieved May 28, 2006, from http://www.tc184-sc4.org/SC4%5FOpen/SC4%5Fand%5FWorking%5FGroups/QC/N-DOCS/maindisp.cfm

TC 184/SC4 Quality Committee. (2001b). *Internal review checklist (QC N193); Project leader checklist (QC N194); and Convener checklist (QC N195).* Retrieved May 29, 2006, from http://www.tc184-sc4.org/SC4%5FOpen/SC4%5Fand%5FWorking%5FGroups/QC/N-DOCS/maindisp.cfm

Thompson, D. (Ed.). (1995). *The concise Oxford dictionary of current English* (9th ed.). Oxford, UK: Clarendon Press.

Warren, T. L. (2005). Communicating style rules to editors of international standards. In *Cross-cultural communication: Perspectives in theory and practice* (pp. 53–70). Amityville, NY: Baywood.

ADDITIONAL RESOURCES

ISO General information:
ISO Central Secretariat
1, rue de Varembê, Case postale 56
CH-1211 Genève 20, Switzerland
TEL: +41 22 749 01 11
FAX: +41 22 733 34 30
E-MAIL: central@iso.org
WEB: http://www.iso.org

About ISO:
www.iso.org/infoe/aboutiso.html

About ISO Technical Committees:
http://www.iso.org/tc

ISO Online:
http://www.iso.ch/

ISO Members:
http://www.iso.org/isomembers

ISO Café:
http://www.iso.org/isocafe

ISO Magazines:

ISO Focus: Overview of standards and standards issues: what is being done and why, how standards affect you. In English. (http://ww.iso.org/isofocus)

ISO Management Systems: Covers ISO 9000 and 14000, the families of standards related to quality. Also covers societal issues related to social responsibilities and the various national initiatives. In English, French, and Spanish. (http://www.iso.org/ims)

SC4 Necessary Documents:
http://www.nist.gov/sc4/www/necsdocs.htm

Global Relevance of ISO Standards:
http://www.tc184-sc4.org/SC4_Open/SC4_Standards_Developers_Info/Files/2004-09-03_-_Global_relevance_of_ISO_technical_work_and_publications.pdf

ANSI Home page:
http://www.ansi.org

NISO Home page:
http://www.niso.org/

CHAPTER 12

Technical Communication in Israel: Training for the Bleeding Edge

Leah Guren

Teaching TC (technical communication) in Israel since 1995 has been a rewarding, yet challenging, experience for me. I have had the privilege of watching the field develop from an almost unknown job to a respected, white-collar technical profession. I have felt great satisfaction at seeing the graduates of my course launch successful careers, many going on to work in, and even manage, tech pubs departments at some of the country's leading high tech companies, such as Kodak, KLA, NetManage, Intel, Kulicke & Soffa, Microsoft, Given Imaging, ECI, and Amdocs.

As technical communication began to attract more attention and become recognized as a desirable profession in Israel, more courses and training programs appeared on the scene. All were commercial ventures; no college or university has, to date, successfully launched a program. Of these commercial programs, some were no more than attempts by untrained (and often unscrupulous) individuals to cash in on the growing market. These questionable commercial courses, which still appear from time to time, usually quickly disappeared, while the better programs have continued to make a serious effort to prepare their students for the real world of TC in Israel. While each commercial TC course taught in Israel differs, the reputable ones have several things in common:

- They are run by real TC practitioners with at least 10 years experience
- They each have a syllabus that matches the needs of the local market
- They each deal with cultural issues that are unique to Israel, such as working with Russian and *Sabra* (native Israeli) SMEs and coping with a seemingly unstructured work hierarchy

Simply put, transporting a technical communication course developed and run in the United States to Israel is *not* a recipe for success. Rather, there are a number of special challenges to teaching technical communication successfully in Israel that would not be met by US-developed TC courses.

This chapter explores these unique challenges from my collaborative work with In Other WORDS, and their course, Introduction to Technical Communication. I initially developed the course in 1995 and have continued to modify it to meet the changing needs of the market. I am both responsible for the syllabus and course material, and am the primary instructor. The course, which runs at In Other WORDS' training centers in Karmiel and Tel Aviv, and the over 250 students who have gone through the program from 1995 through 2009, as well as my own 30+ years in the field as a practitioner, have provided me with the empirical observations of this chapter.

To appreciate the special nature of TC education in Israel, readers must first understand how the following factors affect TC practices:

- The Israeli high tech industry
- How the field of TC in Israel differs from that in much of North America
- How the cultural differences influence the profession
- The makeup of the average candidate accepted into a TC training program

As a result of such factors, TC education in Israel has developed a set of practices that reflects the information needs and expectations of a particular culture.

ISRAELI HIGH TECH

Innovation

Israel may conjure up images of camels against a backdrop of palm trees, but the reality is more Buck Rogers than Indiana Jones. Think *silicon*, not sand.

With one of the highest per capita rates of PhDs in the world and a culture that encourages innovation and discovery, perhaps it is no surprise that Israel pulls in more venture capital money than any other place in the world apart from Silicon Valley (Cnet.com, 2007). Israeli R&D efforts lead the world in chip design, network technology, telecom innovation, and biomed patents (Organization for Economic Co-operation and Development, 2007). Even agricultural life on the kibbutz has contributed to the high tech explosion with automated dairy management systems, solar water heaters, and drip irrigation.

Much has been said about the Israeli ability to innovate. Dr. Geert Hofstede (1997) was one of the first to look at personality traits of societies with his cultural dimensions analysis. While some of Hofstede's work has been questioned in recent years, his initial ideas about what makes a national culture tick remain insightful. Hofstede's emphasis is on understanding cultures to improve sales and marketing efforts, but the basic concepts can be extrapolated to other areas. For example, one clue to Israeli innovation may be connected to the extremely low PDI (Power Distance Index), which measures the range of power (financial or political), or how deeply hierarchical a society is (Hofstede, 1997). A high score shows a society with a large gap between the rich and the poor or between those with political clout and those shut out of the system. An extremely low score indicates a very flat social hierarchy, without large gaps between the *haves* and *have nots* (Hofstede, 1997). In Israel, this is quite visible in high tech, where the only currency of honor or respect is merit, not position. In such an environment, the lowliest programmer has no problem arguing with a top-level manager or challenging the design ideas of a more senior engineer.

Historically, a tradition of questioning developed within Judaism. Rather than rewarding rote memorization, scholarship always included the right of the student to ask questions and engage in discussions. Most of the great rabbis throughout the ages where those who were able to find creative solutions in their interpretation of *halacha* (Jewish law) (Ben-Menahem, Hecht, & Wosner, 2005).

But whatever the root reasons, no one can argue with the statistics. Israel produces more per capita technology patents than most other developed countries, including the United States (Organization for Economic Co-operation and Development, 2007).

International Perspective

This cultural proclivity for innovation, when coupled with a society that embraces technology, creates an interesting mix. More Israelis per capita own home computers than Americans (BigNewsNetwork.com, 2007), and cell phone saturation is one of the highest in the world (Cellular subscribers forecast by country, 2005). Israelis embrace gadgets with none of the trepidation noticeable in other markets. Technology is not viewed as suspicious or geeky, but a legitimate intellectually challenging pursuit and desirable career, attracting almost 8,000 new engineering graduates each year (Cordis, 2007). Israel invests a whopping *five times* more in technology R&D (as a proportion of business R&D) than the United States (Central Bureau of Statistics, 2007). It is no surprise then, that many American high tech companies open R&D centers in Israel. Intel, Cisco, and Microsoft are just a few of the companies that have turned to Israel as a source of advanced technology manpower.

On the flipside, technology companies founded in Israel almost inevitably take advantage of American expertise in marketing, usually by having marketing planning handled through their offices in the United States, and sales and support offices globally. Even with a technologically savvy consumer group like Israel, the actual market size in this country with only 7 million people is tiny. Therefore, virtually every high tech company must market to the world to thrive. Luckily, Israeli technology firms have been doing that successfully for many years. Simply look at the presence of Israeli technology in our modern world: Windows XP and NT operating systems, the media chips in many MP3 players and digital cameras, popular PC CPU chips, and ICQ (instant messaging) are just some of the computer-related technology that the Western world takes for granted—and all based on Israeli patents. Other fields, including agriculture (drip irrigation, new hybrid fruits, water desalinization) and biomed (pill-sized cameras, cardiac stents, laser corrective eye surgery) have also made an impact globally (Israel21c, 2007).

The Emergence of Technical Communication

This need to sell complex technology products outside of Israel made the high tech market one of the first to recognize the importance of technical communication. Suddenly, documentation became more than a "nice to have" add-on; it was a necessity. In some cases, as with biomed companies dependent on FDA approval, documentation became as critical as any other aspect of the development process (US Food and Drug Administration, 2002). Further, this situation meant that product interface and documentation had to be created in English; after all, it is far easier to find a good translator to localize a product from a source language of English than from Hebrew! For the companies who spread their development efforts internationally, all technical documents, including requirements and design specifications, were written in English. But being able to read and write English well enough to communicate with another developer was not the same as being able to write enduser documentation. While, on paper, English is one of Israel's official languages, and most college-educated Israelis have a functional command of English, it is *not* a mainstream language (with less than 2.5% of the population being native English-speakers) (NationMasters.com, 2007). Suddenly, a critical profession emerged that required professional-level English skills in a country where English was not a widely used language.

Israeli Technical Communication Challenges

Unlike their colleagues in North America, Israeli TCs work almost exclusively in high tech industries such as chip design, networking software, and biomed products. They not only must cope with highly complex products, but they must be the resident English experts, often without the luxury of other sources within the organization for editing or support. They are more likely to have broad responsibilities rather than narrow niche roles, and are often involved in document design, production, and graphics. These professional TCs are expected to use a wide range of software applications to get their work done. Additionally, since these TCs are ensconced in a highly dynamic environment, they are expected to cope with constant change and absorb new information as fast as it comes. One graduate of our course told me, "I thought you were exaggerating about the pace of change, but it's worse. I come back from a bathroom break and the programmers have already stuck another version [of the product] on the server!" In this environment, TCs are left trying to document a moving target. Like a circus clown riding a unicycle and juggling, the Israeli TC must do all this with the added complexity of a multilingual (and multicultural) environment.

Rachel J., a TC who worked in Israel for many years before moving back to North America, sees the difference both in the writer's mentality and the way the writer is perceived by the company. "As a tech writer in Israel, I was much more part of the technical team. Here [in North America], tech writers seems less adventurous with new technologies, and the career itself is not prestigious, as [it is] in Israel."

Ahuva C., a senior writer at one of the Israeli branches of a major international high tech company, adds, "The documentation produced by other tech pubs departments [within the company but outside of Israel] is very basic. All the cool stuff—the interactive tutorials, the new generation Help—was produced here. It is as if Israeli technical communicators are less afraid of coding, of figuring out technology, of troubleshooting advanced software applications. No fear."

ISRAELI POPULATION AND CULTURE

Immigration

Since its founding in 1948, Israel has absorbed over 3 million Jewish immigrants from all over the world, including approximately 118,000 from North America (United States and Canada) and another 60,000 from other English-speaking countries (Great Britain, South Africa, Australia, New Zealand). While representing less than 6% of the total immigration and only about 2.5% of the total population, the Anglo community enjoys prominence disproportionate to its size (Haaretz.com, 2007).

For the most part, Anglo immigrants come by choice, not necessity. They are not refugees fleeing persecution, but individuals who are pursuing a dream. As such, they bring with them more financial and academic resources than other groups. Anglo immigrants, particularly from North America, are likely to hold college degrees and be white-collar professionals. This is certainly a good way to start a new life in another country, but there is one serious problem: many professions require recertifying. A lawyer, accountant, architect, or teacher cannot simply go to work without learning a new legal system, a new set of tax laws, a new set of building codes, and a new set of education standards—and all of it in Hebrew. Imagine learning a language well enough to pass a law board exam! For many Anglos, the reality is that they will never master Hebrew well enough to be able to practice their original profession. Even if they can master the language, there is a chance that their profession may not exist in Israel or may not provide enough income.

Others, particularly those who came to Israel when they were in their 20s, ended up with manual labor jobs on a kibbutz. At 22, the spirit of the collective agricultural community seems romantic and idealistic; at 42, getting up at 4:00 a.m. for the first milking makes the promise of an office job quite appealing. Additionally, many of these Anglo immigrants take advantage of their language skills by teaching English in the public schools, a stressful environment that further enhances the appeal of white-collar work.

So whether they are new immigrants who cannot use their existing careers in Israel or well-established immigrants facing burnout from physically demanding or terribly stressful jobs, they form a group of educated adults looking for a new profession that will take advantage of their mother-tongue English skills.

Cultural Adaptation

Anglo immigrants face another challenge in addition to language; coming from cultures where polite behavior is the expected lubricant for most social and business interactions, Anglos often have trouble adjusting to the forthright Israeli approach.

For example, several years ago, past STC (Society for Technical Communication) president Saul Carliner was in Israel speaking to a group of writers, editors, and information architects. He started his lecture by describing a scenario in which the technical communicator, an outside contractor on a project, had been called in to document a project life cycle. He painted a careful picture of a high-level manager who expected a glowing tribute to the brilliant project, and the decidedly bleak perspective of developers who thought that the whole project was an unmitigated fiasco. He challenged the group to strategize on a solution. Most of the senior TCs at the event looked at him as if he had sprouted a second head. Finally, someone asked, "What's the problem? Why didn't the developers just tell the manager that the project was crap?"

Carliner was somewhat taken aback. "You have an interesting culture," he said. That one exchange has come to symbolize to me all that is different between American and Israeli corporate culture. The flat social hierarchy and the relative lack of respect for rank or position makes the Israeli culture ideal for innovation, but also leads to interactions that seem brusque, and even (to an Anglo ear) downright rude.

The Candidate Profile

It is understandable that so many Anglo immigrants want to enter the field of TC. They are often motivated by a need to find an alternative career, especially one that takes advantage of their English skills or the desire to change to a less physically demanding or a less stressful career. This may explain the more mature age (mid-40s) of most candidates entering technical communication courses.

The same is true for other immigrants; while over 85% of those entering TC courses are Anglo, there remain other candidates with good written English skills and a strong aptitude for the profession. These candidates are usually mother-tongue Hebrew (i.e., born and raised in Israel) or Russian (part of the massive influx from the former Soviet Union). But we have also had students who spoke Spanish, German, Dutch, Norwegian, Romanian, Arabic, and French.

About 70% of the candidates are in their 40s, with another 15% over 50 and another 15% under 40. Almost two-thirds hold at least a first college degree. Less than 10% of candidates have not already had a profession. (These are the very young students who have just finished the army or their university degree, but have not yet entered the workforce.)

The respect and reasonably good salary that goes with the profession attract many people, often for the wrong reasons; an entry-level TC in the Tel Aviv area can make NIS 10,000 per month gross (about USD $2,500[1]—more than twice the salary for an entry-level high school teacher), and more seasoned writers can earn upwards of NIS 15,000 (CJI, 2007; Klein, 2007). Adults with no technical aptitude, no computer experience, and no strong written communication skills often think that they can quickly learn enough to start making money. Many have a very sketchy idea of what a technical communicator actually does or how much pressure and responsibility go with the job. Some even think that it is possible to be a dilettante, dabbling at the work for a few hours a day. Others think that this is a field appropriate for introverts, little realizing that the Israeli TC must be extremely aggressive and proactive in gathering information. These myths all influence many people to look to technical communication as a career parachute, when, in fact, it might be decidedly unsuitable for them.

This candidate profile (mature adults looking to retrain for a new profession) affects many aspects of a TC course. These are people who have not been students for many years; they are often rusty at basic student skills (note-taking, staying awake during lectures, etc.). Unlike college-age students, they have many other demands on their time, including families and full-time jobs. They have little patience for theory that is not grounded in *tachlis* (practicality).

[1] Based on the shekel-to-dollar exchange rage of June 2007.

BUILDING A TECHNICAL COMMUNICATION PROGRAM IN ISRAEL

Screening Applicants for Technical Communication Programs

There was once a course in Israel that became notorious for its ads gushing: "If you are a mother-tongue English speaker, you can be a technical writer!" How offensive! To me, that is like saying, "If you have steady hands, you can be a brain surgeon!" There is just so much more to it! The huge amount of knowledge; the extensive theory; the logical analysis; the visual skills; the communication aspect; the mastery of tools. Needless to say, not everyone is cut out to be a good TC, and it is unethical for any training program, whether a private course or a university path, to accept anyone based solely on the language of their country of origin and the solvency of their checking account. Happily, that particular program has long-since closed its doors, but from time to time other TC-wannabe programs pop up, only to ultimately flounder.

In Other WORDS' course was the first to start screening candidates. In the following years, other courses came to understand that this was the best way to prevent disappointment and failure from candidates who were not suitable for the profession. Not only does screening help weed out the technophobes and other obvious bad fits, but it allows the instructors to maintain the pace and intensity needed to cover the material in the syllabus.

In Other WORDS' screening process is in the form of a 2–3 hour meeting with no cost or obligation on the candidates' part. We do our best to make sure that people have enough understanding to make an informed decision and to screen people for basic skills and aptitudes. The process includes

- Briefly introducing the company and the background of the program.
- Explaining the realities of TC in Israel. This is our chance to address many of the myths and misconceptions that people have about the field while painting a realistic picture of the pluses and minuses of working in high tech. It is amusing to note that many candidates lose interest when they discover that TC here requires long hours, strong technical aptitude, and close interaction with other people. Others come in with no idea that the field is so technical.
- Explaining the realities of the course. Because of economics, no one can afford to take a year or two off to retrain. Thus, our course is an intense 5-month program, conducted like a graduate-level program. For every hour of lecture time, students can expect an additional 2–3 hours of preparation time, between reading, homework assignments, and practicing. There are tests and a very demanding final project. Many candidates are surprised to learn that the course will be difficult and time-consuming, and choose to opt out.

- Testing candidates on basic computer skills.
- Administering a written exam to look for writing skills, editing ability, knowledge of grammar and punctuation, logic, math, and technical knowledge.
- Interviewing each candidate individually to get some sense of their personality, background, and previous work experience.

This entire process serves two important functions: first, it allows us to identify the best possible candidates and to eliminate the obviously unsuitable, while giving us enough subjective data (the personal interview) to deal with those who fall into the grey area. Secondly, it provides an opportunity for candidates to ask questions and make sure that there are no surprises, either about the profession or the course itself. Candidates are then contacted and informed whether or not they have been accepted into the program.

While the screening process is not perfect, it does help prevent many people from investing in the time, money, and energy in a professional training program that is ill-suited for their skills or temperament. Screening remains the only option for a commercial course that wishes to behave ethically toward students, by not accepting candidates who have no realistic chance to succeed in the field.

Providing Precourse Preparation

Without the luxury of a long-term program, participants are expected to work on certain skill areas before the course begins. This includes a review of English grammar and punctuation, as well as some general technical material (such as a review of technology terms, basics of computer viruses, and other background information).

Most students are also required to take a prerequisite course in Windows and technology. This course (also taught by In Other WORDS) covers such basics as a conceptual understanding of hardware components, software and operating system overviews, Web research skills, and Windows "power user" best practices. This is a jump-start to allow them to catch up with other students who have had more hands-on computer experience. Not only is the content important, but the process helps get students into the correct mindset for being proactive about research and information. As discussed in "The Emergence of Technical Communication" (earlier in this chapter), few writers here have the luxury of having a formal editor, so the English review is crucial. Though this is also covered in the class, the head start is necessary for most students who either never learned proper punctuation and grammar or for the nonnative English speakers who learned a mishmash of rules from an international assortment of teachers. Without this precourse, we would waste too much time during the course proper, and students lacking this background would quickly flounder. In structuring a commercial course, there is a delicate balance between covering critical information and keeping the course short enough to be manageable for students and cost-effective for the company.

Pushing a Theory-Based Syllabus

I have always believed that a good TC course syllabus must stress theory and its application to the work, rather than just tools. By theory, I mean the logic and rationale behind best practices, rather than an academic philosophy. For example, *know your audience* is a piece of theory; learning how to capture audience data and perform audience needs-analysis and task-analysis is the practical extension of that theory. As for tools, while students must master the basics of DTP tools (desktop publishing), HATs (Help authoring tools), graphics tools, and HTML editors, it is more important that they learn proper work habits and techniques than rote memorization of how to do something. As the tools change (think of the massive changes in the UI from version to version of some of the more well-known DTP tools, for example), good work habits still apply. Similarly, TC theory remains applicable no matter how much the products we document, the tools we use to document them, and even the final deliverables, change.

In addition to the lecture syllabus, the supporting reading material ideally covers a wide range of views and perspectives. Since little in this field is black and white, it is important that students quickly learn how to cope with conflicting viewpoints and seemingly contradictory information.

To be able to sort through the volumes of information, students need to develop critical thinking. Unfortunately, this is nearly impossible to teach. (This is one reason why we screen for logic skills; candidates with poor logic often fail to develop the ability to think logically and analyze information, or to be able to make good decisions by extrapolating existing data.) Sadly, these aptitudes are set at a much younger age, and all that a good course can do is further hone them. This seems to be one of the deciding factors of a graduate's success in the field: those individuals who are uncomfortable with ambiguity and want to memorize a set group of rules to apply never seem to be able to thrive, while those who can look at a situation, analyze the information, and make a decision by applying the theory dynamically, do much better.

Several years ago, I was called in by a client to evaluate their existing documentation suite and make recommendations for their next release. I met with the tech pubs department, where three of the writers, including the department manager, were graduates of my course. While reviewing one of the manuals, I noticed something jarring in the layout and asked the group why they had done that. They explained the process they went through to balance the needs of the audience, the aesthetics of the presentation, and the technical limitations of the single-sourcing tools they were using, and how they ultimately arrived at that particular decision. Listening to them, I felt enormous pride at their professionalism. They had assimilated the theory, mastered analysis, and were making tough decisions to balance ideals with pragmatism.

From an instructor's standpoint, I find that while my underlying theory concepts have stayed consistent over the years, the practical application, the examples used, and the homework assignments that exercise that theory continue to evolve. Recognizing the building blocks of practical TC theory and applying them in an up-to-date method that reflects the realities of the field is crucial. Again, as our students are studying with an immediate career in mind, they need to master the thought processes that will allow them to continue to learn and grow on the job.

Creating a Business Environment

A training course that operates like an academic program is doomed to failure, especially if modeled on Israeli higher education practices, where deadlines are never final and pressure is relative. Unless someone is studying TC just from intellectual curiosity, a "normal" academic program won't serve any purpose. It certainly won't help a student cope with the realities of the high tech market or the expectations of a tech pubs manager.

Therefore, our course establishes tough business guidelines from the beginning, where the instructors take on the dual role of SMEs (subject matter experts) and clients. Students are given strict guidelines about the mechanics of submitting assignments, and deviations are rejected. Deadlines are always fixed, not vague moving targets that can be missed without impacting a student's grade. Most students find this very difficult—even painful—but by the end of the course, they have developed a work method that more closely matches the rhythm of high tech than academia.

"I thought your rules about deadlines . . . were unnecessarily harsh," former student D. told me, "but that's what it's really like out here. I was surprised how prepared I felt when I started working with real clients."

Course participants who have made careers out of being students find the transition unpleasant. "You find something wrong with every assignment," complained one student who was used to getting straight As in a college environment.

"Welcome to the real world," responded the instructor dryly.

Students who came from other professional careers adapt to these demands much more easily. They usually have a better understanding of the concept of providing service to clients or customers and are more thick-skinned about making changes to documents. They can separate their ego or personal identity from the assignment, and focus on applying the theory they learned to the task at hand.

TC instructors in Israel should be able to help students extrapolate from their previous work experience, using analogies and comparisons to shorten the learning curve.

Adding Cultural Conditioning

Students who have lived in Israel for more than a few years are usually quite familiar with the mentality and mannerisms of the people they will work with. New immigrants, however, must wrestle with a completely different set of rules about social interactions, business expectations, and an endless supply of potential communication pitfalls. Much of this is something that can only be learned by going through the process of *klitah* (absorption into Israeli society), but there are things that we have added into the course to help the process:

- Understanding the perspective of developers: A good TC needs to deal with grey areas, while the average developer tends to reside in a world of binary thinking. Students have to learn to consider the SMEs' thought process so that they can correctly translate information. Learning to interview SMEs by asking scenario-based questions (open, not closed, questions) helps keep students from missing critical information. Students also need to learn how to create and maintain good relations with SMEs, by being an English resource, by learning to read code, by organizing their questions to minimize annoying or repetitive interruptions.
- Developing a Hebrew technical vocabulary: The writing is in English, but meetings, water-cooler discussions, and project gossip are not. Students need to learn a few key terms in Hebrew, along with a robust English technical vocabulary, to help them "talk the talk and walk the walk" with the developers.
- Working with management: The flat hierarchy of business interactions is hard for some nonnative students to adjust to. They expect managers to offer direction and are uncomfortable with the idea that employees must sometimes initiate change or direct a project. The idea that they have to be very aggressive to get the information needed at the start of a project, and to set certain guidelines, such as rules for reviews during the edit cycle, is alien. We try to break them of the habit of passively accepting a project at face value by giving several assignments and pop quizzes without enough information. Students eventually learn to question and define things before starting.
- Role playing: Role playing is a useful tool and can help students mentally prepare for certain tasks, such as interviewing. Nonnative students are often shocked by the seemingly rude (i.e., personal) questions that come up during the average job interview.

Would any course in the United States have to teach students to interpret hand signals and body language from a culturally alien group of SMEs? The irony is that the very characteristics that make Israelis excel in technology R&D are the things that are most alien to Anglo students. It is not enough to teach someone to read an RC file or hack a CSS; a successful TC in Israel must learn interpersonal communication skills that seem at times counterintuitive.

SUMMARY: THE X FACTOR

What Makes Teaching TC in Israel Different?

In conclusion, TC in Israel is a vibrant, well-established profession that is well-recognized and equally well-respected. Teaching TC is rewarding and demanding, and the differences can be summarized as follows:

- Technology is king: A highly technical market, with the vast majority of the TC jobs existing in high tech, requires that any TC course provide a strong focus on technology. Some courses have included elements of programming and electronics to make their students more "literate."
- Students are retooling: The unique nature of our students, with a pool of mature, educated white-collar professionals who are entering a second (or third) career, means that these are individuals who may have been out of an academic environment for over 20 years. They need to brush off rusty skills, such as note-taking and researching, and they often need to adjust to the rigors of frontal (lecture mode) training. On the other hand, they bring with them a certain business savvy, professionalism, work ethic, and maturity. Most of them see the course as a serious financial investment and understand that they will get more out of it if they put more into it.
- Screening is essential: While many people think that this is a good way to make money, they are ill-suited for the demands of the profession here. Candidates must be screened for language skills, logic, and technical knowledge and aptitude. In addition, they must be fully counseled about the true nature of the profession to prevent anyone from making a decision based on myths and inaccuracies.
- The real world is tough: A traditional academic environment won't prepare students here for what they will face as in-house writers or contractors. Deadlines, submission rules, attention to detail, and an early indoctrination into working with SMEs and clients is crucial to allow students to survive.
- Theory is the cornerstone: Hands-on instruction only makes sense when students are applying solid theory. Things like audience analysis, signposting, and design based on user cognitive responses have to be covered in detail, with the students then applying them in realistic writing assignments.
- Cultural adaptation helps success: Students must be made aware of the differences in social cultures and how they affect work, projects, and interactions. More than talking about it, role-play and interactive exercises can help students prepare.

CONCLUSION: LESSONS LEARNED

What Do Teachers of Technical Communication in Israel Need to Know?

Teaching a successful course in Israel requires

- Experience as a working TC professional in high tech, not as an academic.
- Reasonable fluency in Hebrew, even if all course material is in English.
- A willingness to abandon topics that aren't useful here, such as P&P (policies and procedures).
- Sensitivity toward the needs of an older student, long out of school, busy with a family and job, and wanting efficiency.
- The ability to create work habits (such as rigid adherence to deadlines) that reflect the reality of the field.

What Do Students of Technical Communication in Israel Need to Know?

Potential students need to consider the following:

- It's a technical field. If you can't imagine yourself sitting in on geeky engineering meetings about software design, this may not be for you.
- Be prepared to unlearn and relearn a lot of English.
- Don't bank on stability. High tech is up and down, with companies adding staff only to downsize the next quarter. If you are uncomfortable with change, this may not be for you.
- If you are a big-picture person, you'll be challenged with the details, and vice versa. If you are left-brain, you'll be challenged with right-brain tasks, and vice versa. In short, be prepared to leave your comfort zone.
- Bring your curiosity, logic, and people skills.

By understanding and examining these factors, Israeli TC educators can better address workplace needs and develop classes and curricula that address such needs. TC educators in other nations, in turn, can use the ideas covered in this chapter to better prepare their students to interact with Israeli counterparts in future global projects. Through such increased understanding, all involved parties can provide students with the knowledge base and skillsets needed to succeed in the global workplace of tomorrow.

REFERENCES

Ben-Menahem, H., Hecht, N. S., & Wosner, S. (2005). *Controversy and dialogue in the Jewish trader.* London: Routledge.

BigNewsNetwork.com. (2007). Israeli tech breakthroughs to increase speed of computers 10-fold. Retrieved November 13, 2007, from http://feeds.bignewsnetwork.com/?sid=18699

Central Bureau of Statistics. (2007). *CBS, statistical abstract of Israel 2007.* Retrieved November 10, 2007, from http://www1.cbs.gov.il/shnaton58/st28_15.pdf

CJI – Documents/Resources. (2007). Computer jobs in Israel. Retrieved November 12, 2007, from http://www.cji.co.il/docs.htm

Cnet.com. (2007). *The world's tech hot spots.* Living with technology: Tech without borders. Retrieved November 12, 2007, from http://www.cnet.com/4520-13653_1-6794290-2.html?tag=nav

Cordis. (2007). *Israel's R&D capacity: A promising land.* Retrieved November 11, 2007, from http://cordis.europa.eu/israel/rd_en.html

eTForcasts. (2005). *Cellular subscriber forecast by country.* Retrieved November 13, 2007, from http://www.etforecasts.com/products/ES_cellular.htm#1.1

Haaretz.com. (2007). *AngloFile briefs.* Retrieved November 11, 2007, from http://www.haaretz.com/hasen/spages/919619.html

Hofstede, G. (1997). *Cultures and organizations: Software of the mind.* New York: McGraw-Hill.

Israel21c. (2007). Retrieved November 11, 2007, from http://www.israel21c.org/bin/en.jsp?enPage=HomePage

Klein, Z. (2007). *Striking teachers to receive lower salaries.* Globes online. Retrieved November 12, 2007, from http://www.globes.co.il/serveen/globes/DocView.asp?did=1000269878&fid=1725

NationMasters.com. (2007). *Language statistics: English speakers (most recent) by country.* Retrieved November 11, 2007, from http://www.nationmaster.com/graph/lan_eng_spe-language-english-speakers

Organization for Economic Co-operation and Development (OECD). (2007). *Compendium of patent statistics.* Retrieved November 12, 2007, from http://www.oecd.org/dataoecd/5/19/37569377.pdf

U.S. Food and Drug Administration. (2002). *General principles of software validation: Final guidance for industry and FDA staff.* Retrieved November 11, 2007, from http://www.fda.gov/cdrh/comp/guidance/938.html#_Toc517237933

CHAPTER 13

An Overview of Technical Communication Education in New Zealand

Richard Draper

The New Zealand economy is changing rapidly into a "knowledge economy," which requires a more highly skilled workforce. With the movement to higher skills comes the need for accurate dissemination of complex and technical information—the realm of the technical communicator. Additionally, New Zealand's relatively small economy makes the nation highly reliant on trade with other countries, and such trade increasingly involves cross-cultural communication.

This chapter will examine New Zealand and the changing nature of its economy, the need for technical communication in New Zealand's high tech industries, and the training or education of technical communicators to meet such needs. In particular, this chapter will cover the following items:

- New Zealand and its economy, with particular interest in the movement of the economy toward higher technology.
- Technical communication in high technology organization, illustrating some of the issues and solutions to cross-cultural technical communication.
- Advent of the profession of technical communication and professional organizations, showing how the arrival of professional organizations helped establish the profession of technical communication.
- Training in technical communication in New Zealand, with a brief look at training courses.

Through this examination, readers will gain a better understanding of technical communication in New Zealand and how professionals are trained to enter jobs in that field. From this examination, readers can gain a better understanding of both the nature of technical communication in New Zealand and how to train technical communication students in other nations to work with counterparts in New Zealand.

NEW ZEALAND'S PEOPLE

New Zealand is a small South Pacific island nation, famed for its natural beauty, its "clean, green" reputation, and its sporting successes (out of proportion to its size). More recently, New Zealand has gained fame for its role as a backdrop for movies such as *The Lord of the Rings* and *The Chronicles of Narnia: The Lion, the Witch and the Wardrobe*. Roughly the size of the United Kingdom and Japan, New Zealand has only just reached a population of 4 million: 4,027,947 according to the 2006 census (Statistics New Zealand, 2007). This population, however, is very diverse, and such diversity affects a variety of practices, including those related to technical communication.

Although a member of the British Commonwealth, New Zealand is a cultural mix of European (68%), Mâori (14%), Asian (9%), and Pasifika (7%) (Statistics New Zealand, 2007).

Interestingly, only 11% of the overall population identify themselves as New Zealanders, an indication of an ongoing debate on the national and cultural identity of the nation (10% of the population identify themselves as having more than one ethnic background) (Statistics New Zealand, 2007). And while New Zealanders tend to be understated in their attitudes on culture and identity, their dedication to their native land has earned New Zealand the enduring nickname of "Godzone" for "God's own country" (Johnstone, 1995, para. 1).

NEW ZEALAND'S ECONOMY

New Zealand's relatively small size and its isolation have enabled some interesting social and economic "experiments" to be conducted, from the "Welfare State" of the 1930s (see, for example, Baker, 1965) to the sweeping reforms of the 1980s and 1990s. These latter years saw New Zealand's economy, and society, change from one largely dependent on primary produce (exports of sheep and dairy products) to an open economy, with the removal of both tariffs and restrictions on foreign ownership of key assets and infrastructure.

More recently, the government of New Zealand has been aiming to create a "knowledge economy" that would produce higher technology goods and services for the rest of the world. The following World Bank statement illustrates the need for such an economy:

> For countries in the vanguard of the world economy, the balance between knowledge and resources has shifted so far toward the former that knowledge

> has become perhaps the most important factor determining the standard of living—more than land, than tools, than labour. Today's most technologically advanced economies are truly knowledge-based. (World Bank, 1999, as cited in Ernst & Young, 1999)

The move toward a knowledge economy is not to say that primary production is no longer important. Indeed, tourism (based on natural resources) is increasing in importance to the overall economy. However, more than research and innovation are required to produce goods and services relevant to and desired by greater international markets and that have significant value added *before* being exported.

To address such factors, organizations in New Zealand have invested heavily in research as varied as varieties of fruit (for example, long-lasting and more differently flavored apples; see ENZA, 2007), more efficient home appliances, and cutting edge motion picture and television technology. The company Fisher & Paykel, for example, has developed electronic technology to power and operate home appliances in a way that would lead to smaller and more convenient devices (Fisher & Paykel, 2007). At the same time, companies such as Weta Workshop and Weta Digital have gained international renown for their innovative visual effects such as those seen in *The Lord of the Rings* trilogy of movies (Weta Workshop, 2007).

There is also a growing software development presence in New Zealand. The Jade Software Corporation, for example, uses its own software development platform to create both Jade software products and a number of products produced through the use of Jade software. The efficient, object-oriented nature of the platform, moreover, enables small numbers of software engineers to develop complex computer systems in a relatively short time and has allowed the Jade Software Corporation to produce a container port management software product, Jade Master Terminal, with a small team of fewer than 10 programmers. (In comparison, Jade's leading competitor in the world market uses a variety of database and programming languages and has a staff of over 300 (see Navis, 2007)).

Part of New Zealand's success in these high-tech areas can be attributed to the traditionally high cost of technical equipment, such as computers, and the relatively low buying power of the New Zealand dollar. These constraints, in turn, are credited with prompting many New Zealanders to try to wring as much value out of their equipment as possible and to find new, more self-reliant ways of producing technology products.

Also contributing to this success is the fact that New Zealand's relatively small and isolated economy, in combination with its people's reputation as early adopters of technology, has made New Zealand an ideal location for companies to trial run new technology products. The Electronic Funds Transfer at Point of Sale (EFTPOS) financial network, for example, was adopted quickly in New Zealand when it was introduced in 1985, and New Zealand now has "more EFTPOS terminals per head of population than any other country, and being used for about 60% of all retail transactions. According to the largest EFTPOS network provider, 'New Zealanders use EFTPOS twice as much as any other country'" (Paymark, 2009; *Wikipedia*, 2007). These factors and developments, in turn, gave rise to the need for professional technical communicators to operate in New Zealand's growing high-tech sectors.

TECHNICAL COMMUNICATION IN NEW ZEALAND

In the early 1980s, the New Zealand-based LINC Development Centre created a software product—the Logic and Information Network Compiler (LINC)—for a multinational organization with customers worldwide. As part of the delivery of this software product, technical documentation on the use, operation, and administration of the product was required. In response to this demand, a technical communication group was formed as part of the LINC Development Centre. This technical communication group originally delivered a paper (printed) version of each document in its product information library. This paper was in turn shipped to Unisys Corporation offices in the United States for reproduction and shipment with the LINC software. At that time, documentation was produced on mainframe computers using a text editor that was based on a simple form of single-sourcing with conditional tagging.

Over time, the business world changed its expectations of computer organizations from being suppliers of hardware and software products to suppliers of "business solutions." As a high productivity development tool, LINC thus became more and more important to the U.S.-based Unisys Corporation. This development, in turn, had implications for the technical communication group of the LINC Development Centre in terms of both the preparation of documentation.

Documentation was eventually moved to Unisys B20 computers with PC-like technology and a more sophisticated tagging scheme. These changes enabled WYSIWYG preparation and delivery of product information/documentation as electronic files (rather than paper) directly to Unisys publication units and to translation centers. In response to this change, the LINC library was moved to SGML in order to enable the delivery of paper, online, and CD versions of product documentation to an organization's library from a single source.

Over time, the size and complexity of the LINC software grew, and it was delivered on a wider range of mainframe platforms, each of which had its own requirements. The size of the product information library grew too and eventually reached 5,000 and 10,000 pages in length, including associated training courses. The library was delivered as updates two to four times a year and in major releases every 1 to 2 years, often in parallel (major releases and updates in overlapping timeframes) by a group of around a dozen technical communicators. This task was no easy feat, for in addition to its size and technical detail, the library also needed to be delivered to customers in a variety of countries and cultures. This factor meant that cross-cultural communication became an integral part of technical communication practices in New Zealand.

CROSS-CULTURAL TECHNICAL COMMUNICATION

To cater to a worldwide market, Unisys established translation (localization) centers in the United States, Asia, and Europe. It also researched cross-cultural communication and developed guidelines for the writing of product information for a global market. As LINC employees were tied into these processes at a very intimate level, their documentation activities became intertwined with Unisys's translation and localization efforts. It was intended that as much translation as possible be done in the translation centers using translation software, and the task completed by translators, in an attempt to make the task as cheap and fast as possible.

To assist with this objective, technical communicators were given a set of guidelines that would help them provide materials better suited for quick and effective translation. These guidelines can be reduced to the following six principles:

- Use simple English, with active (second person) voice and the smallest appropriate words. For example, instead of "dialogue about the utilization of" use "talk about the use of." (See, for example, Plain English, 2007.)
- Use consistent language. Unlike a novel, where the varied use of language can add interest, technical communication is intended to be clear and unambiguous. Repeating information in the same way in similar contexts (for example, using the same stem sentence), adds to confidence and readability, rather than detracting linguistically from the communication. This principle applies also to headings and to document layout (Johnson-Sheehan, 2002, p. 197).
- Reduce the overall amount of material as far as possible. This generally involves being clear about the purpose of the communication and being committed to that purpose. In the software industry, this has been known to cause conflict between technical communicators and software engineers, with the latter often keen to have documented *how* a particular function is engineered as well as what it does. From a practical point of view, reducing the quantity of material to be translated reduces time, cost, and opportunity for error in the translation.
- Provide a comprehensive glossary that defines jargon and other important terms.
- Avoid colloquialisms, idiomatic expressions and the like, as they usually do not translate well. For example, the expression "the spirit was willing, but the flesh was weak" was fed into English-to-Russian translation software

and the result fed into Russian to English software. The result was "the vodka was good, but the meat was rotten."

- Replace Latin constructions with simple English: for example, *and so on* for *etc; that is,* for *i.e.; for example* for *e.g.*

These guidelines are, in fact, sound principles for effective technical communication in general, and they can also be seen in the minimalist movement (for example, Eiler, 1997) and in commercial approaches to technical communication. The Information Mapping (Information Mapping, 2007) approach to technical writing, for example, is based around clearly identifying types of information, separating these, and ensuring that they are clearly labeled.

These six initial principles, however, represent just the beginning of cross-cultural communication. That is, even a simple, minimalist approach might not always be appropriate. In fact, what might be considered simple and direct in one culture could be viewed as rude and abrupt in another (for example, "high context" cultures in which meaning is derived as much from the context of the situation as it is from the actual words used in that situation (Eunson 2005, pp. 467–468)). These factors, in turn, proved to be of great importance as New Zealand's technical communicators became more involved with international activities.

During the late 1980s, for example, the training courses associated with the LINC software products needed to be updated. These courses were developed to help Unisys train its software engineers in showing customers how to make the best use of the LINC software. One of the courses developed was a 2-week administration course, which was a mixture of how to administer and maintain the LINC software and how to design a software system for development using LINC. It was decided to base the design issues on a well-known organization, Domino's Pizza, in order to enable a wide range of issues to be covered. For example, inventory control, delivery, customer service, and financials. In addition, it was felt that Domino's Pizza would be familiar to customers in the United States, the largest portion of the Unisys customer base.

The first hour of the course was an introduction and a quick overview of the kind of service offered by Domino's Pizza, including "If your pizza is not delivered in 30 minutes, it's free!" This approach seemed to work well until one of the U.S.-based Unisys instructors reported back. He had been sent to one of the Baltic states, which were, at that stage, still part of the Soviet Union.

As it happened, when the Baltic-based instructor started the administration course, he noted that his audience sat quietly though the introduction, but looked puzzled. When he asked for questions, they asked about how the organization worked because they did not understand. "Here," they said, "if you order a pizza, there is only one flavor. And it will be ready in two weeks." The unconscious underlying assumption in this course was that all attendees would have at least an understanding of a capitalist economy! As a result, the New

Zealand-based technical communicators who developed the related training materials needed to begin considering thinking of their audiences on a much more global scale. A valuable resource in achieving this objective was the rise of professional organizations dedicated to technical communication.

THE ADVENT OF THE TECHNICAL COMMUNICATION PROFESSION AND PROFESSIONAL ORGANIZATIONS

In the early 1990s, the Unisys LINC Development Centre was expanding rapidly. As a result, the organization had formed a documentation group dedicated to creating its product information and courses as well as attend to the qualification (testing) of its software products. Outside of Unisys LINC, however, technical communication, if considered at all, was still widely regarded as part of software development. Thus a relatively poorly qualified and poorly motivated software engineer, rather than a trained technical communicator, often wrote most technical documentation.

This situation is not uncommon within the field of technical communication. Rather, Hackos (1994, p. 46) describes a model of organizational maturity that characterizes levels of organization to its treatment of, attitude toward, and processes around technical communication. At lower levels, technical communication (if needed at all) is seen as a near unnecessary chore, to be completed on a whim (usually last). At higher levels, processes are managed, and technical communication actually adds value to products. This model is outlined in Table 1.

Within this framework, the technical communication field in New Zealand has tended to hover at levels 0 and 1. In fact, even today, with the rise of a variety of high tech businesses in the country, this attitude continues in some parts of the software industry.

A study of the course prescriptions from the National Advisory Committee on Computing Qualifications (NACCQ) of New Zealand illustrates this current attitude toward technical communication. Founded in 1988 as a result of a review of the national qualifications in data processing and information technology, NACCQ specifies prescriptions for some of the courses run in the tertiary education in New Zealand, and it is made up largely of representatives from industry and academia (see NACCQ, 2007). As a result, NACCQ has a great deal of authority over technical and professional communication education in New Zealand. Moreover, the perceptions NACCQ representatives have of what constitutes technical and professional communication shape the nature of courses in those areas. For example, in the "BC500 Business Communication" course overseen by NACCQ, the skills and knowledge students are to gain from such a course include

Table 1.

Level	Label	Description
0	Oblivious	Unaware of the need for professional publications. Produced by whoever is available. No project management.
1	Ad Hoc	Technical communicators work independently to produce publications with little or no coordination. Often assigned to different technical managers. No project management.
2	Rudimentary	Some coordination between technical communicators for consistency, but enforcement is not strong. No or very little project management.
3	Organized and Repeatable	Sound development process in place. Some project management in place.
4	Managed and Sustainable	Strong project management in place to ensure the publication development works. Estimating and tracking of projects is thorough. Innovation gains importance in strong existing structure. Strong commitment to project management.
5	Optimizing	Everyone engaged in monitoring and controlling projects. Self-managed teams become the norm. Strong commitment to continuous improvement, to project management, and to instituting self-managed teams.

1. Types and characteristics of written communication
2. Simple research process
3. Effective writing skills
4. Technical writing (NACCQ, n.d.)

In terms of technical writing education, the outline of NACCQ-related technical writing course/educational objectives consists of

Technical Writing

- The characteristics of effective technical documents used in the information technology environment are gathered and considered. The range must include manuals, online help, procedures documentation, and electronic content.

- Design an effective technical document. Points to consider include font, color, layout, sections/structure, sequence, graphics, labeling, index, audience, language, sentence structure, jargon levels, navigation, and such.
- Write a technical document for a specified ICT [information and communication technology] audience.
- Evaluate the document produced against preset standards. This should include some reference to user testing. (NACCQ, n.d.)

These factors indicate NACCQ's (and presumably significant parts of the software industry's) opinion is that technical communication is a function of software engineers. Fortunately, members of the software industry in New Zealand generally do not share this opinion.

PROFESSIONAL ORGANIZATIONS AND TECHNICAL COMMUNICATION EDUCATION IN NEW ZEALAND

While the Unisys LINC Development Centre was expanding rapidly in the early 1990s, the organization found it difficult to recruit experienced and trained technical communicators. This problem was due in large part to the fact that there was little recognition of the profession and few training courses offering formal training in the field. This situation also provided much-needed motivation for developing ties with the Society for Technical Communication (STC) (STC, 2007a) in order to sponsor the creation of a full STC chapter in New Zealand (STC NZ, 2007). The STC New Zealand chapter, in turn, sought to encourage the establishment of technical communication courses in New Zealand.

These developments were not entirely altruistic gestures, as the STC New Zealand chapter also sought to offer professional technical communication services to other high technology companies. The idea was that doing so would heighten recognition of the technical communication/technical writing profession and would, in turn, help members profit from such an association.

The goals of the overall STC organization are to

- Enhance the professionalism of the members and the status of the profession;
- Provide information through publications, reports, and conferences;
- Report on new communication technologies, methods, and applications;
- Provide recognition and awards;
- Provide services to members at all levels of the Society;
- Promote the education of members and support research activities in the field;
- Give service to industry and academe. (STC, 2007b)

The New Zealand chapter of STC appears to have succeeded in attaining these goals. That is, the profession of technical communication there appears to have gained recognition. For example, the Open Polytechnic, a tertiary institution in New Zealand, recently advertised a certificate course in technical communication in response to a demand from industry. As the related press announcement noted, "Responding to industry demand, The Open Polytechnic of New Zealand has recently launched . . . the Certificate in Technical Communication" (Open Polytechnic, 2006, level 5).

In addition, another institution, Unitec, indicates that the profession (career) of technical communication is recognized as such: "the Certificate in Technical Communication equips you with the skills for a career as a technical writer" (Unitec, n.d.).

Today the membership of STC in New Zealand has decreased for a number of reasons, including its cost; the U.S.-centric nature of the STC; and the advent of a New Zealand-based organization, the Technical Communicators Association of New Zealand (TCANZ), which addresses the niche once filled by the STC. In 1997, members of STC New Zealand created TCANZ which actively continues to encourage and promote the profession of technical communication in New Zealand (TCANZ, 2007). These factors, have shaped and continue to shape technical communication education in New Zealand.

TECHNICAL COMMUNICATION COURSES IN NEW ZEALAND

Having moved to establish the profession of technical communication in New Zealand, the Unisys LINC Development Centre sought to establish some specific training courses. The group first approached various tertiary institutions, and a positive response was received from the Christchurch Polytechnic Institute of Technology (CPIT). As a result, CPIT worked with local industry to develop a technical communication program, the Diploma of Professional and Technical Communication. Within the program's curriculum, the concepts of audience, document design, and project management were seen as central.

In 1995, the whole program was first offered, and was done so online—a brave innovation given the relative state of Internet communication software at the time. Lack of reliability and slow speed of connections, the newness of the medium, and consequent lack of student exposure to online education were problematic at first. However, by 2007, CPIT had managed to resolve most of these issues.

Over time, the initial CPIT program has evolved into the Graduate Diploma of Technical Communication and the Graduate Diploma of Information Design (GDID).

Within this educational framework, the GDID exists as a 1-year, full-time program that is delivered entirely online. A graduate diploma from this program is roughly equivalent to the first half of a master's degree program at conventional universities, and prospective students are required to possess a bachelors degree or equivalent before they can enter the program. In fact, the GDID may be used toward gaining a Master of Science in Information Design and Communication through the Southern Polytechnic State University (SPSU) in Marietta, Georgia (CPIT, 2007).

The GDID has been very successful and continues to attract online students from both New Zealand and around the world. The relative value of the New Zealand dollar to the U.S. dollar, moreover, has proven beneficial in comparison, for it is relatively inexpensive for a U.S. student to gain the first half of his or her master's degree in technical communication from the CPIT. In fact, anecdotal evidence suggests that some students consider the cost so cheap that it could not be a "real" qualification!

The highly popular GDID program consists of the following courses:

- Writing and editing
- Research and theory in communication (includes cross-cultural communication)
- Analysis and management; includes project management
- Document design
- Usability testing
- Internship; industry experience (CPIT, 2005)

Interestingly, the degree title of Graduate Diploma of Technical Communication was recently changed to Graduate Diploma of Information Design. This change from "technical communication" to "information design" perhaps reflects a debate in the technical communication profession worldwide. It perhaps also reflects the changing nature of the profession itself as it moves away from simple writers of technical material, to "visual and verbal language structures" (Schriver, 1997, p. 6) responsible for planning and delivery of information for many types of output (paper, CD-ROM, online help, Web pages, and so on). Schriver (1997, pp. 1–11), for example, devotes a chapter to the subject of defining the field. She notes the various terms used and appears to favor "communication design" (p. 8). She also notes its shortcomings (p. 9).

This debate is still continuing in the greater international technical communication community. According to the International Council for Technical Communication (INTECOM), technical communicators are also known as documentation developers or technical writers, and more recently the titles information designers or information architects have been proposed. We have taken the term "technical communicator" as covering what we do in the broadest sense because the term technical writer does not fully describe the scope of our work in which design and architecture are an integral part. However, the debate is not closed and we will continue to examine what we do and ensure that what we call ourselves represents our role (INTECOM, 2005, para. 2).

It is certainly a brave move for CPTI to choose the title Graduate Diploma of Information Design instead of retaining the commonly used title of the profession (technical communication). For example, in a search of "technical communication course" in New Zealand using Google, a link to the GDID program appears 15th on the list, and then apparently only by accident (as the reference occurs in an endorsement from a student who uses the previous name "Graduate Diploma of Technical Communication").

The courses in the GDID appear to have good coverage of the requirements of a technical communicator, from writer through designer to usability test and manager. The GDID makes no attempt to cover software applications, although this is understandable in an asynchronous, online environment. In addition, knowledge of specific software applications is likely to be as transient as the applications themselves.

As an aside, it is interesting to note that even though the program is delivered entirely online, details of the courses in the GDID are not included on the CPIT Web site. In fact, the GDID program does not appear to be connected to any other CPIT qualification, although, as previously noted, it does offer a pathway to a Master of Science in Information Design and Communication (CPIT, 2007).

Another technical communication program offered in New Zealand is the Certificate in Technical Communication (Unitec, 2006). This certificate consists of the following courses:

- Digital Communication Applications: software skills for technical communication
- Technical writing: writing skills for technical communication
- Developing On-line [sic] & Web documentation: theory and practice for creating online documentation

Taking online or on-site classes can complete the Certificate in Technical Communication and is part of the set of Communication programs listed below, which feature considerable emphasis on international and intercultural communication (Unitec, 2007a):

- Bachelor in International Communication
- Postgraduate Diploma in International Communication
- Master of International Communication

The courses included in the curriculum for the Certificate in Technical Communication do cover writing and have an emphasis on visual and graphic information. However, such courses do not cover all of the skills Schriver (1997) considers required by technical communicators and that are considered key by

INTECOM (2005). (Usability testing and documentation project management, for example, are not included in the curriculum for the Certificate. Interestingly, the Certificate in Technical Communication, unlike the GDID, does offer training in using software applications.)

Another institution in New Zealand, the Open Polytechnic (an institution that also specializes in distance learning), offers a Certificate in Technical Communication, which consists of a compulsory course and one of two other courses (Open Polytechnic, 2007; Unitec, 2007b):

- Writing for specific purposes (compulsory): plain English writing, presentation, and editing, plus specialized options (for example, writing for law)
- Professional and technical writing: effective business writing
- Writing for the Web: principles of writing for Internet use

When combined with other courses at Open Polytechnic, this qualification can lead to the completion of a Bachelor of Arts or a Bachelor of Applied Science with a major in Communication (Open Polytechnic, 2007).

Although the courses in this Certificate in Technical Communication do include editing and document preparation, overall, there does not appear to be coverage of the skills required by technical communicators, as discussed in Schriver (1997) and noted by INTECOM (2005). For example, Schriver (1997), in her discussion of the name of the profession, goes further by noting "From the perspective of document design, however, the most serious problem with 'technical communication' or 'professional communication' is that these rubrics fail to suggest that writers must be able to think visually as well as verbally" (p. 10).

This Certificate in Technical Communication, by contrast, does appear to give the impression of minimizing the profession of technical communication.

While other institutions in New Zealand offer communication courses, none of them appears to offer specific technical communication courses. It is also interesting to note the emphasis institutions such as Unitec (Unitec, 2007a) have on communication—it is an international and intercultural approach to largely *corporate* communications and may be a pointer to where the technical communication profession is headed.

SUMMARY

This chapter has briefly described New Zealand, its economy, and the role of technical communication in that economy. The establishment of the profession of technical communication and the development of training courses has been illustrated, and one of these courses, the Graduate Diploma of Information Design, offers training to students worldwide.

It is interesting to attempt to speculate on future requirements for technical communication, even if that name continues to be relevant or accurate to the requirements of the profession. In a world where more and more individuals have a voice and there appears to be an increasing emphasis on the way in which information is presented rather than its content (see, for example, McLuhan & Fiore, 1967), there appears both threat and opportunity for the profession.

REFERENCES

Baker, J. (1965). *War economy.* (Part of *The Official History of New Zealand in the Second World War 1939–1945.*) Wellington: Historical Publications Branch. Retrieved February 1, 2007, from http://www.nzetc.org/tm/scholarly/tei-WH2Econ-c1-4.html

CPIT. (2005). *Graduate diploma in information design* [Brochure]. New Zealand: Author.

CPIT. (2007). *MSc in information design.* Retrieved March 14, 2007, from http://www.cpit.ac.nz/subjects/informationdesign/msc_information_design.php

Eiler, M. (1997). Minimalism and documentation downsizing: The issues and the debate. *The Newsletter of the Chicago Chapter of the Society of Technical Communication, 39*(4).

ENZA. (2007). *Various varieties [of apple].* Retrieved March 6, 2007, from http://www.enzafoods.co.nz/information/various_varieties/

Ernst & Young. (1999). *The knowledge economy: A submission to the New Zealand government by the minister for information technology's IT advisory group.* Retrieved February 27, 2007, from http://www.med.govt.nz/pbt/infotech/knowledge_economy/index.html

Fisher & Paykel. (2007). *Double dishdrawer.* Retrieved March 3, 2007, from http://usa.fisherpaykel.com/global/kitchen-products/dishwashing/dishdrawer.cfm?productUid=31173CB5-65B7-DB96-15435FCB74BA3BA3&image=insit&section=

Hackos, J. (1994). *Managing your documentation projects.* New York: John Wiley & Sons.

INTECOM. (2005). *Technical communication as a profession.* Retrieved March 13, 2007, from http://www.intecom.org/index.php?option=com_content&task=view&id=12&Itemid=32&lang=EN

Johnson-Sheehan, R. (2002). *Writing proposals: Rhetoric for managing change.* New York: Longman.

Johnstone, B. (1995). Godzone. *WIRED.* Retrieved November 10, 2009, from http://www.wired.com/wired/archive/3.11/newzealand.html

McLuhan, M., & Fiore, Q. (1967). *The medium is the massage.* London: Penguin.

NACCQ. (2007). *Fostering computer education in New Zealand.* Retrieved March 20, 2007, from http://www.naccq.co.nz/index.html

NACCQ. (n.d.). *BC500 Business communication.* Retrieved February 20, 2007, from http://www.naccq.co.nz/bb_view/bbpre.html?view=yes&tbl2=BC500

Navis. (2007). *Navis corporate background.* Retrieved March 5, 2007, from http://www.navis.com/about.jsp

Open Polytechnic. (2006). *Technical communication certificate launched.* Retrieved March 12, 2007, from http://www.openpolytechnic.ac.nz/aboutus/newsandevents/technical-communication-certificate-mr

Open Polytechnic. (2007). *Certificate in technical communication.* Retrieved March 14, 2007, from http://www.openpolytechnic.ac.nz/programmesandcourses/ programmes/ op5505

Paymark. (2009). *EFTPOS—Introduction.* Retrieved November 10, 2009, from https:// www.paymark.co.nz/cms_display.php?st=1&sn=37&pg=228

Plain English. (2007). *Fighting for crystal-clear communication since 1979.* Retrieved March 11, 2007, from http://www.plainenglish.co.uk/

Schriver, K. (1997). *Dynamics in document design.* New York: John Wiley & Sons.

Statistics New Zealand. (2007). *New Zealand's 2006 census of population and dwellings.* Retrieved March 1, 2007, from http://www.stats.govt.nz/NR/rdonlyres/6C178AC5-4530-4979-A9DA-96ACD17700B7/0/quickstatsaboutpopulationmobility.pdf

STC. (2007a). [Home page.] Retrieved on March 2, 2007 from http://www.stc.org/

STC. (2007b). *About STC, Vision & Strategic Plan.* Retrieved on March 2, 2007 from http://www.stc.org/about/

STC NZ. (2007). [Home page.] Retrieved on March 2, 2007, from http://www.stcnz.org.nz/home/

TCANZ. (2007).[Home page.] Retrieved March 2, 2007, from http://www.tcanz.org.nz/

Unitec. (n.d.) *Certificate in technical communication.* Retrieved March 12, 2007, from http://www.unitec.ac.nz/?9FA615F2-6C30-4D72-AD16-CED502166F90

Unitec. (2006). *Certificate in technical communication.* Retrieved on March 12, 2007, from http://www.unitec.ac.nz/toolbox_items/bro_certTechComm.pdf

Unitec. (2007a). *Communication studies.* Retrieved on March 14, 2007, from http://www.unitec.ac.nz/toolbox_items/bro_communication.pdf

Unitec. (2007b). *Developing on-line & Web documentation.* Retrieved March 15, 2007, from http://www.unitec.ac.nz/toolbox_items/bro_communication.pdf

Weta Workshop. (2007). [Home page.] Retrieved March 7, 2007, from http://www.wetaworkshop.co.nz/ and http://www.wetafx.co.nz/digital/

Wikipedia. (2007). *Debit card.* Retrieved March 20, 2007, from http://en.wikipedia.org/wiki/Debit_card

World Bank. (1999, as cited in Ernst & Young, 1999). *World development report 1998/99: Knowledge for development.* Washington, DC: World Bank.

Contributors

EDITOR BIOS

Kirk St.Amant is an associate professor of Technical and Professional Communication at East Carolina University. He has a background in cultural anthropology, international government, and technical communication, and his research interests include intercultural communication via online media. He has taught online and conventional courses in technical and professional communication and in intercultural communication for East Carolina University, Texas Tech University, Mercer University, and James Madison University. He has also taught courses in e-commerce, distance education, and business communication in Ukraine as a part of the USAID-sponsored Consortium for the Enhancement of Ukrainian Management Education (CEUME) and engaged in collaborative online teaching with the Aarhus School of Business in Denmark.

Barry Thatcher is an associate professor of Rhetoric and Professional Communication at New Mexico State University. His research interests include intercultural professional communication, intercultural rhetorical theory, history of rhetoric in Latin America, and intercultural research methods. He is also an adjunct professor at La Universidad Autónoma de Ciudad Juárez and on the advisory board of the Colegio de Chihuahua, both of which are public universities in Mexico. He is the founder of Border Writing, a research and testing firm for cross-border technology integration and training. His most recent projects include testing new GPS and RFID technologies for securing cross-border hazmat shipments and using ESL classrooms for health literacy instruction.

CONTRIBUTING AUTHOR BIOS

Deborah C. Andrews directs the Center for Material Culture Studies at the University of Delaware (www.udel.edu/materialculture) and coordinates the English Department's concentration in professional writing. She has published several texts on professional communication, including *Technical Communication in the Global Community* (2nd ed., Prentice Hall, 2001) and, most recently, *Management Communication: A Guide* (with W. D. Andrews: Houghton Mifflin, 2004). The former editor of *Business Communication Quarterly,* she has contributed articles to major journals and anthologies in communication, most recently, "North America, the State of the Field," a chapter in the *Handbook of Business Discourse* (Edinburgh UP, 2009).

Yvonne Cleary is a lecturer in technical communication at the University of Limerick, Limerick, Ireland, where she is Program Director for the Graduate Certificate/Diploma/MA in Technical Communication. She teaches courses on the theory and practice of technical communication. Her research interests include international communication, virtual teams, and the Irish technical communication context.

Richard Draper is a senior lecturer at Christchurch Polytechnic Institute of Technology (CPIT) in Communication. He completed his PhD in the early 1980s, researching the use of computers for teaching. Richard is a senior member of the Society for Technical Communication (STC) and is a founding member of and current president of the New Zealand Chapter of STC. In addition, he is a corporate member of the Technical Communicator's Association of New Zealand (TCANZ). Prior to his time at CPIT, Richard spent more than 20 years in technical communication, mainly in the software industry. He won an award in the STC International Competition for technical marketing manuals.

Dacia Dressen-Hammouda is associate professor of English for Specific Purposes at Université Blaise Pascal (Clermont-Ferrand, France). She teaches in the department of Applied Foreign Languages for Business and Technology, where she heads the master's program in Languages and Technical Communication. Her research interests include document design, situated genre analysis, English for academic and professional purposes, L1 and L2 writing, and French and English contrastive rhetoric.

Lyn F. Gattis is assistant professor of Professional Writing in the Department of English at Missouri State University. She teaches beginning and advanced undergraduate courses in technical writing and graduate-level document design. She is currently researching linguistic patterns that improve the comprehensibility of single-sourced text. Her other research interests include document design, project management, and history of technology.

Leah Guren is the owner/operator of Cow TC. She has been active in the field of technical communication since 1980 as a writer, manager, Help author, editor, consultant, and trainer. She now devotes her time to consulting and training. Her clients include many of the leading high tech companies in Israel and Europe. Leah is also well-known as a popular speaker at international conferences in the field. She is completing a 3-year term (May 2007 to May 2010) on the board of directors of STC (Society for Technical Communication).

Brent Henze is associate professor of English at East Carolina University, where he directs the internship program. His scholarship on the rhetoric of science, reporting genres in ethnological science, and the scientific treatment of racial difference has appeared in *Technical Communication*, *Technical Communication Quarterly*, and *Rhetoric Review*. He is also co-author (with Jack Selzer and Wendy Sharer) of *1977: A Cultural Moment in Composition* (Parlor Press, 2008).

Charles Kostelnick is a professor in the English Department at Iowa State University where he has taught business and technical communication and a graduate and an undergraduate course in visual communication in professional writing. He was the co-author (with Michael Hassett) of *Shaping Information: The Rhetoric of Visual Conventions* and co-author (with David Roberts) of *Designing Visual Language: Strategies for Professional Communicators.*

James Melton is an assistant professor of Business Communication at Central Michigan University. His research focuses on intercultural communication and second language issues in professional contexts, including offshore outsourcing, sustainable business practices, and global information systems. He is particularly interested in how research findings in these areas can inform professional communication curriculum and pedagogy.

Poornima Padmanabhan has an MA in Educational Technology and Learning Design from Simon Fraser University, Canada. She has worked as a senior technical communicator for several years in both Canada and India, and conducted/published research in instructional technologies, online learning, and technology for development. Her research interests include intercultural communication, social studies of information technology, and information design. Poornima's most recent project was a social and educational study of the One Laptop Per Child (OLPC) initiative in India.

Makarand (Mak) Pandit is a technical communicator, academician, and businessman with over 15 years of experience. He is one of the pioneers of technical writing in India. His company, Technowrites, serves top Indian IT and engineering companies, and Global Fortune 500 companies. Mak was the owner of Technical Writers of India mailing list for over 5 years. He is a senior member of the Society for Technical Communication and has served the society in many volunteer positions (including as a president of India Chapter in 2005). Mak teaches at many educational institutes including University of Pune. He can be reached at mp@vsnl.com.

Shelley L. Smith is currently an associate professor in the Instructional Development Service at the University of Minnesota, Duluth. The focus of her 25 years in academia has been in intercultural and adult and faculty education; internationalizing the on-campus curriculum; and programming and training in international, educational, corporate, and community contexts.

Angela Thevenot teaches in the Applied Linguistics program at the University of Memphis. She has also taught in EFL programs in the Czech Republic and Mexico; presented at regional, national, and international conferences; and currently conducts ESL workshops for teachers in the greater Memphis area. Her research interests include English for Academic Purposes, EAP student interface with the academic discourse community, second language acquisition and critical pedagogy, and grammar anxiety in the language classroom.

Emily A. Thrush teaches in both the Applied Linguistics and Professional Writing programs at the University of Memphis. She has been an Academic Specialist for the USIA/State Department in the Brazil, Czech Republic, Lebanon, Slovakia, Italy, and Germany and served as a Senior Fulbright Fellow in Mexico in 2000–2001. She is co-author of several books in the McGraw-Hill.

Thomas L. Warren, Professor Emeritus of English and former Director of the Oklahoma State University Technical Writing Program, has taught technical writing since 1972 in universities and industry in the United States and Europe. He developed the BA, MA, and PhD technical writing degree programs at OSU and is Guest Professor at the University of Paderborn (Germany), teaching Report Writing, Critical Reading of Technical Texts, and Presenting at Conferences to undergraduates, graduate students, and faculty. Past president of INTECOM, he is both an ATTW and STC Fellow, recipient of CPTSC's Distinguished Service Award, and STC's Jay R. Gould Award for Excellence in Teaching.

Index